Disclaimer

The publisher of this book is by no way associated with the National Institute of Standards and Technology (NIST). The NIST did not publish this book. It was published by 50 page publications under the public domain license.

50 Page Publications.

Book Title: A Welding Data Dictionary

Book Author: William G. Rippey;

Book Abstract: This data dictionary specifies data elements that are used to describe the design, fabrication and inspection of weldments. It comprises welding information from several authoritative sources. Data include description of component materials, details of joints and welds, and variables of the welding processes. The data dictionary provides, in one document, the welding information needed to develop format specifications for digital data that can be exchanged by computer programs.

Citation: NIST Interagency/Internal Report (NISTIR) - 7107

Keyword: data dictionary;digital welding data;Standards;weld inspection results;Welding;welding terms

NISTIR 7107

A Welding Data Dictionary

William G. Rippey
U. S. DEPARTMENT OF COMMERCE
Technology Administration
National Institute of Standards
 and Technology
Intelligent Systems Division
Gaithersburg, MD 20899-8230

National Institute of Standards
and Technology
Technology Administration
U.S. Department of Commerce

NISTIR 7107

A Welding Data Dictionary

William G. Rippey
U. S. DEPARTMENT OF COMMERCE
Technology Administration
National Institute of Standards
 and Technology
Intelligent Systems Division
Gaithersburg, MD 20899-8230

March 2004

U.S. DEPARTMENT OF COMMERCE
Donald L. Evans, Secretary

TECHNOLOGY ADMINISTRATION
Phillip J. Bond, Under Secretary for Technology

**NATIONAL INSTITUTE OF STANDARDS
AND TECHNOLOGY**
Arden L. Bement, Jr., Director

A Welding Data Dictionary

Abstract

This data dictionary specifies data elements that are used to describe the design, fabrication and inspection of weldments. It comprises welding information from several authoritative sources. Data include description of component materials, details of joints and welds, and variables of the welding processes. The data dictionary provides, in one document, the welding information needed to develop format specifications for digital data that can be exchanged by computer programs.

Keywords

data dictionary, digital welding data, weld inspection results, welding procedure specification, welding terms, weldment design.

Foreword

In 1992, welding information formats were described in two American Welding Society (AWS) specifications, *AWS A9.1- Standard Guide for Describing Arc Welds in Computerized Material Property and Nondestructive Examination Databases*, and *AWS A9.2 - Standard Guide for Recording Arc Weld Material Property and Nondestructive Examination Data in Computerized Databases*. This document expands the scope of those documents and could serve as a core document if an AWS committee undertakes a new specification project. Current individual AWS specifications are rich in terms and data items describing materials, design and processes. This data dictionary gathers the information in a single document, to be used by data modelers with specific encoding formats in mind. Examples of format specifications are STEP and XML. The ultimate purpose is to facilitate exchange of welding data between application software programs developed by different vendors.

TABLE OF CONTENTS

Abstract i
Keywords i
Foreword i
TABLE OF CONTENTS ii
Scope 1
Purpose 1
   Source Documents 2
   Guidelines Used in Building the Data Dictionary 2
   Sections of this Document 3
   Dictionary Field Definitions 3
1. Weld Design Data 5
   1.1 element base_metal 6
   1.2 element connection 9
   1.3 type edge_shapes 11
   1.4 element joint_design 11
   1.5 element joint_penetration 15
   1.6 type joint_types 16
   1.7 type linear_dimension_range_type 17
   1.8 type linear_dimension_type_mm 18
   1.9 element sub_assembly 18
   1.10 element weld_details 19
   1.11 element weld_sizes 20
   1.12 element weldment 22
   1.13 element workpiece 22
   1.14 type weld_types 22
2. Welding Process Data 23
   2.1 type allied_processes 23
   2.2 type arc_welding_processes 23
   2.3 type brazing_processes 23
   2.4 element electrical_specification 24
   2.5 element filler_metal 27
   2.6 element gas_component 30
   2.7 type gas_flowrate_type 30
   2.8 element heat_treatment 31
   2.9 type other_welding_processes 33
   2.10 type oxyfuel_gas_welding_processes 33
   2.11 element preheat_and_interpass 34
   2.12 type resistance_welding_processes 35
   2.13 element shielding_gas_for_procedure 35
   2.14 type shielding_gas_type 38
   2.15 type soldering_processes 39
   2.16 type solid_state_welding_processes 39
   2.17 type temperature_type 39
   2.18 element variables_specification 41
   2.19 element welding_position 48

  2.20 element welding_procedure_specification 49
  2.21 type welding_process_type 50
3. Testing and Inspection Data 52
  3.1 element bend_test_bend_radius 52
  3.2 type bend_test_types 53
  3.3 element fracture_crack_plane_orientation 53
  3.4 element fracture_energy_absorbed 53
  3.5 element fracture_machine_notch_position 54
  3.6 element fracture_specimen_location 54
  3.7 type fracture_toughness_test_methods 54
  3.8 type fracture_type_of_test_equipment 54
  3.9 element hardness_indentor 55
  3.10 element hardness_load 55
  3.11 element hardness_location_of_impressions 55
  3.12 element hardness_test_result 55
  3.13 element hardness_test_types 55
  3.14 element mechanical_test 56
  3.15 type mechanical_tests 57
  3.16 element nick_break_apparatus 58
  3.17 type shear_specimen_types 58
  3.18 element shear_test_shear_strength 58
  3.19 element shear_test_unit_shear_load 59
  3.20 type soundness_test_types 59
  3.21 type tension_specimen_types 59
  3.22 element tension_test_data 60
  3.23 element test_fracture_appearance 62
  3.24 element test_fracture_location 62
  3.25 element test_LGBTFW_angle_of_fracture 63
  3.26 element test_LGBTFW_discontinuity 63
  3.27 element test_machine_serial_number 64
  3.28 element test_number_of_specimens 64
  3.29 element test_percent_elongation 64
  3.30 element test_postweld_mechanical_treatment 64
  3.31 element test_postweld_thermal_treatment 64
  3.32 element test_specimen_dimensions 64
  3.33 element test_specimen_location 65
  3.34 element test_specimen_orientation 65
  3.35 element test_technician_name 65
  3.36 type test_temperature 66
Acknowledgments 66
Appendices 66
  Appendix 1 – Items to Add in Next Edition of Data Dictionary 66
  Appendix 2 - Tasks for data modelers 67
  Appendix 3 – The BIG PICTURE of welding data exchange 68
  Appendix 4 – Terminology differences between ISO and AWS 69

A Welding Data Dictionary

Scope
This document includes data describing design of welded products, design of weldments including detailed weld and joint description, arc welding process specification, and destructive weld inspection. Common welding documents that convey this information are welding procedure specifications (WPSs) and mechanical testing reports. These documents are often implemented using hard copy or simple, but non-standard data formats. Variables of other welding processes such as gas and laser welding have not been covered at this time.

Some of the variables and data types needed in standard welding procedure specifications (SWPSs) and procedure qualification records (PQRs) are not addressed. Examples are ranges of allowed values (e.g., minimum and maximum allowed current), conditions that are expressly not allowed, conditions that are optional, or multiple conditions that are allowed. These should be straightforward extensions of the data elements and types already defined.

Also not addressed are overall formats for standard forms or standard test result reports. These are left to data modelers according to the requirements of their customers. Appendix 1 lists many other items that are not included in this document. They are suggestions to committees of experts who could expand the document's scope.

Purpose
A data dictionary is the starting point for developing specifications for data formats for digital welding data. It gathers relevant data in one place, establishes standard labels and descriptions of the items, and designates data types to them. The purpose of a standard data format is to enable independent computer programs (applications) to exchange information with minimal or no cost and effort.

This document's form is suitable for a formal standard to be used by many different manufacturing industries that use welding. The existence of a single data dictionary as source for data modeling will ease the challenge of gathering the welding knowledge, and make implementations of different formats more consistent. Example formats are STEP and XML. Consistency eases the challenge of converting between the differing formats should it be necessary. Easy conversion between formats makes the decision of which format to use less critical. When data modelers discover improvements to the model the changes can be reflected in the high level data dictionary as well as to the implementation being worked on at the time. This way other implementations can benefit from the improvement.

It is envisioned that welding standards documents that contain forms will use this data model to build their own structures to represent the form information. This will enable encoding of the information in a standard format so it may be conveyed electronically and not be limited to hardcopy or proprietary formats.

The data dictionary is written in English rather than a formal modeling language so that:
- welding experts can understand the information and work on it to improve it without learning a computer language.

- it can serve as the common ground between welding experts and the data modelers who will develop detailed specifications.
- it is not biased toward any particular encoding scheme.

This dictionary goes beyond presenting all elementary information units as a flat list. It suggests gathering of elements into frequently used groupings. These groupings are not rigid: data modelers or refiners of the data dictionary can ungroup or regroup as they see fit. The groupings present the information in recognizable units that make understanding the document easier.

This data dictionary is not sufficient to specify the design of a database or of a complete specification for standard data. Its terms and type definitions are the building blocks that need to be supplemented by encoding rules, and relationships like those described in database tables.

Source Documents

These documents contain definitions of the variables, or state the need for variables in report documents to describe welding products, processes or inspection results.
1. AASHTO/AWS D1.5M/D1.5:2002 – *Bridge Welding Code*.
2. ANSI/AWS A9.1-92, *Standard Guide for Describing Arc Welds in Computerized Material Property and Nondestructive Examination Databases*.
3. ANSI/AWS A9.2-92, *Standard Guide for Recording Arc Weld Material Property and Nondestructive Examination Data in Computerized Databases*.
4. ANSI/AWS A3.0: 2001, *Standard Welding Terms and Definitions*.
5. AWS B1.10:1999, *Guide for the Non-destructive Examination of Welds*.
6. AWS B2.1:2000, *Specification for Welding Procedure and Performance Qualification*.
7. AWS B4.0M:2000, *Standard Methods for Mechanical Testing of Welds*. Parts A, B and C were used. Parts D and E are not covered here.
8. AWS D1.1/D1.1M: 2002 *Structural Welding Code – Steel*. See especially Annex E, Sample Welding Forms.
9. Jefferson's Welding Encyclopedia, 18th Edition, American Welding Society, 1997.

Guidelines Used in Building the Data Dictionary
- The primary audience for this document is domain experts in welding. They must be able to understand its content and be able to refine and expand it, without needing computer programming experience. The document then becomes a bridge between them and computer experts, who will encode the data and manipulate it in computer programs that serve the welding experts.
- The item descriptions in this document should not be cited as formal definitions of terms. The descriptions are original, or are shortened or paraphrased versions of formal definitions. The source documents, usually AWS in origin, should be used.
- Sources of most definitions are AWS documents.
- When values are shown as "enumeration of:", this means that only the listed values are allowed. This is a restriction on the allowed values for the element.
- This data dictionary specifies, for some elements, hierarchical relationships of sub-elements, and enumerated values. Element grouping and enumeration are constraints on how the data can be used in a computer program or in a digital file. The constraints

promote increased consistency between different vendors' implementations of products or of digital format specifications.
- The element naming convention of lower case words separated by underscores was chosen to be human readable, and mildly suggesting future tags to be used in formal encoding. The convention is not intended to suggest a preference of one encoding method over another.
- Some provisions are made for specifying units for measured variables, especially to distinguish between U.S. customary, or "inch-pound", and International System of Units (SI) usage. More examination of the need and suggestions for values are needed from welding experts.

Sections of this Document

Weld Design Data describes the desired weld that joins two workpieces. It includes descriptions of the workpieces (e.g., thickness, material), the geometric relationship between the workpieces (e.g., tee, butt), the edge shapes of the workpieces to form a joint, and the desired properties of the weld, such as size and degree of penetration. There are usually several different ways to produce a designed weld, e.g., by varying ***Welding Process Data*** such as welding process, number of passes, thickness of consumable, gas composition, etc. ***Testing and Inspection Data*** is information gathered by measuring characteristics of the weld after it is made, usually to compare to the design data.

The appendices contain next-step information for data modelers, some high level background on the concept of standard data formats for welding, and suggestions for expansion of the scope of the next version of this data dictionary.

Dictionary Field Definitions

- There are two kinds of dictionary entries, "element" and "type". Types are data templates that are reused in multiple elements. The prefix "p:" means primitive, one of Boolean, string, integer or decimal. An element is a unique chunk of information used in describing welding. The distinction between types and elements will be further refined in detailed data modeling – the major criterion is to make the model concise and understandable.
- Diagram - When an element has subelements, a sketch of its sub-elements is shown to aid understanding. Diagrams show only one level of decomposition. Appendix 4 contains a few selected diagrams of multilevel diagrams. In the diagrams, the 3 short bars in some boxes is an artifact of the documentation tool and is not relevant. The small arrow in the lower left corner indicates that the element is defined globally, i.e. it is independent of the element definition it appears in.
- Type – enumeration is used when there is one choice from a list of simple values. *Choice of* is used when there are several non-simple elements, one of which must be picked.
- Values – this field appears only for enumerated types, showing the allowed values.
- Children – list of elements that are parts of this element.
- Used by – the name of elements that contain this element.
- Source – citation of a standard or industry handbook that lists the need for the information. When the source for a sub-element is the same as its parent element, the source information is not repeated.

- Description – prose description, or standard definition if one exists, to allow data modelers to select the appropriate elements for specific applications, forms and reports. It is an important task of welding experts, not data modelers, to develop precise descriptions.
- Underlined type and element names are links that can be used in electronic documents for easy jumping to definitions.

The index below is included for use in electronic documents that can use the links to the definitions. The types and elements are arranged in alphabetic order in each section.

Elements
base_metal
bend_test_bend_radius
connection
electrical_specification
filler_metal
fracture_crack_plane_orientation
fracture_energy_absorbed
fracture_machine_notch_position
fracture_specimen_location
gas_component
hardness_indentor
hardness_load
hardness_location_of_impressions
hardness_test_result
hardness_test_types
heat_treatment
joint_design
joint_penetration
mechanical_test
nick_break_apparatus
preheat_and_interpass
shear_test_shear_strength
shear_test_unit_shear_load
shielding_gas_for_procedure
sub_assembly
tension_test_data
test_fracture_appearance
test_fracture_location
test_LGBTFW_angle_of_fracture
test_LGBTFW_discontinuity
test_machine_serial_number
test_number_of_specimens
test_percent_elongation
test_postweld_mechanical_treatment
test_postweld_thermal_treatment
test_specimen_dimensions
test_specimen_location
test_specimen_orientation
test_technician_name
variables_specification
weld_details
weld_sizes

Types
allied_processes
arc_welding_processes
bend_test_types
brazing_processes
edge_shapes
fracture_toughness_test_methods
fracture_type_of_test_equipment
gas_flowrate_type
joint_types
linear_dimension_range_type
linear_dimension_type_mm
mechanical_tests
other_welding_processes
oxyfuel_gas_welding_processes
resistance_welding_processes
shear_specimen_types
shielding_gas_type
soldering_processes
solid_state_welding_processes
soundness_test_types
temperature_type
tension_specimen_types
test_temperature
weld_types
welding_process_type

welding_position
welding_procedure_specification
weldment
workpiece

1. Weld Design Data

1.1 element base_metal

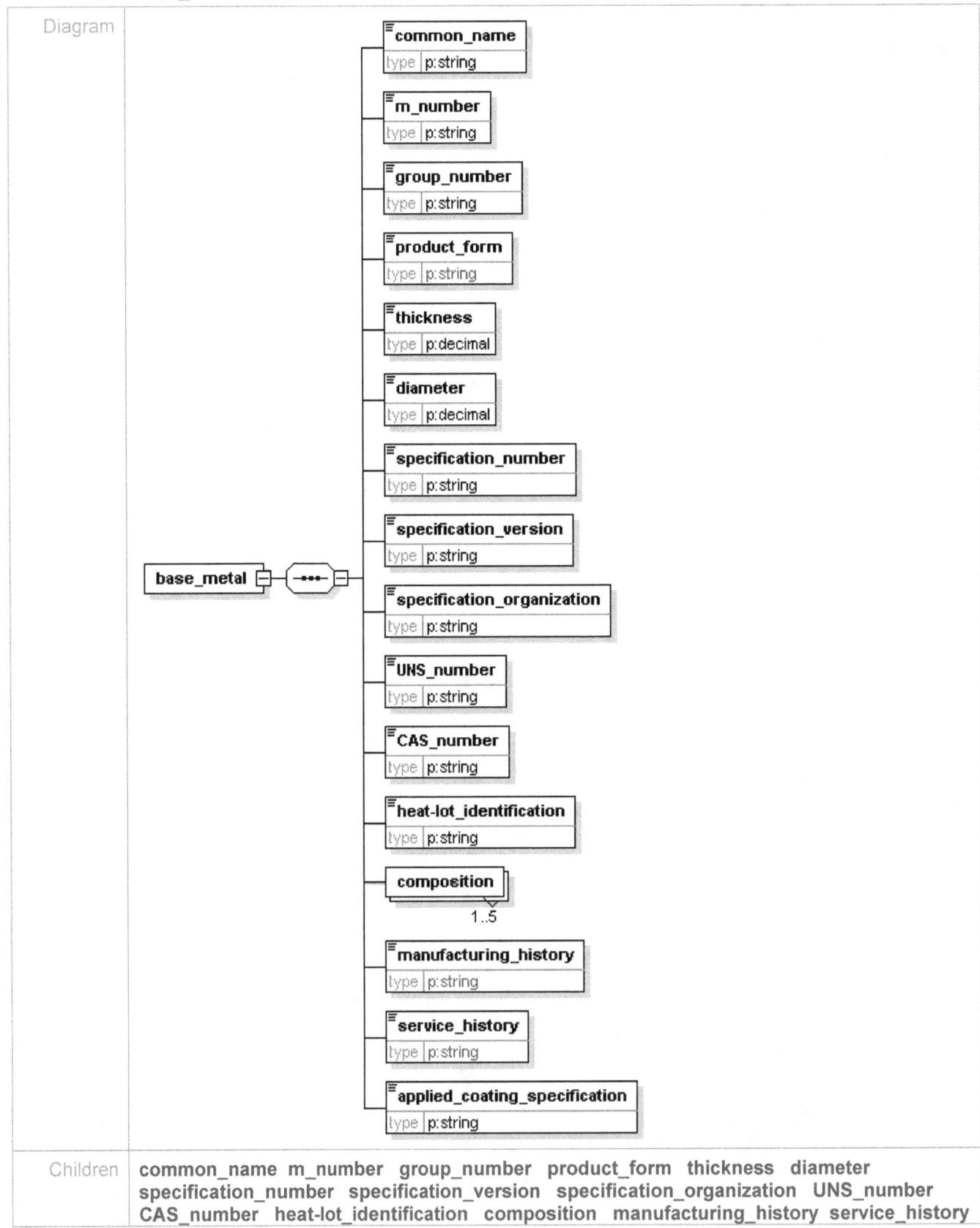

Diagram	
Children	common_name m_number group_number product_form thickness diameter specification_number specification_version specification_organization UNS_number CAS_number heat-lot_identification composition manufacturing_history service_history

	applied_coating_specification
Used By	element welding_procedure_specification
Source	AWS B2.1, section 2.13.
Description	The material that is being joined using a weld.

1.1.1 element base_metal/*common_name*

Type	p:string
Description	The trade name, a name used without all the designations of the formal specification.

1.1.2 element base_metal/*m_number*

Type	p:string
Source	AWS A9., section 5.1.1., and B2.1, glossary definition p. 2.
Description	A designation used to group base metals for procedure and performance qualifications.

1.1.3 element base_metal/*group_number*

Type	p:string
Description	A classification system for metal by material properties.

1.1.4 element base_metal/*product_form*

Type	enumeration of p:string	
Values	enumeration	plate
	enumeration	bar
	enumeration	sheet
	enumeration	tube
	enumeration	pipe
	enumeration	structural - i beam
	enumeration	structural - other
Description	The form of the workpieces to be joined.	

1.1.5 element base_metal/*thickness*

Type	p:decimal
Description	Plate or sheet thickness, if tube, specifies wall thickness.

1.1.6 element base_metal/*diameter*

Type	p:decimal
Description	Outside diameter, only used if material is tube.

1.1.7 element base_metal/**specification_number**

Type	p:string
Source	AWS A9.1, section 5.1.1
Description	The standard designation of the formal material classification.

1.1.8 element base_metal/**specification_version**

Type	p:string
Source	AWS A9.1, section 5.1.1
Description	Version of the specification used.

1.1.9 element base_metal/**specification_organization**

Type	p:string
Source	AWS A9.1, section 5.1.1
Description	The organization responsible for generating the specification.

1.1.10 element base_metal/**UNS_number**

Type	p:string
Source	AWS A9.1, section 5.1.1
Description	Unified Numbering System for Metals and Alloys, managed by ASTM and SAE.

1.1.11 element base_metal/**CAS_number**

Type	p:string
Source	AWS A9.1, section 5.1.1
Description	Chemical Abstracts Service Registry Number, a unique identifier for substances issued by the Chemical Abstracts Service.

1.1.12 element base_metal/**heat-lot_identification**

Type	p:string
Source	AWS A9.1, section 5.1.1.
Description	A unique identifier issued by a materials manufacturer assigned to manufacturing batches.

1.1.13 element base_metal/**composition**

Description	Detailed chemical composition, by elements. This Type needs expansion.
Source	AWS A9.1, section 5.1.1.10

1.1.14 element base_metal/*manufacturing_history*

Type	enumeration of p:string
Values	enumeration cold worked enumeration normalized enumeration annealed
Source	AWS A9.1, section 5.1.1.11.1
Description	Mechanical manufacturing methods used to produce the welded material.

1.1.15 element base_metal/*service_history*

Type	enumeration of p:string
Values	enumeration sour crude enumeration need expansion here
Source	AWS A9.1, section 5.1.1.11.1
Description	The mechanical forming and heat treatment methods used to produce the stock material.

1.1.16 element base_metal/*applied_coating_specification*

Type	p:string
Source	Applies especially to AWS B4.0, section B2-9(1)
Description	Standard designation for the class of coating.

1.2 element **connection**

Diagram	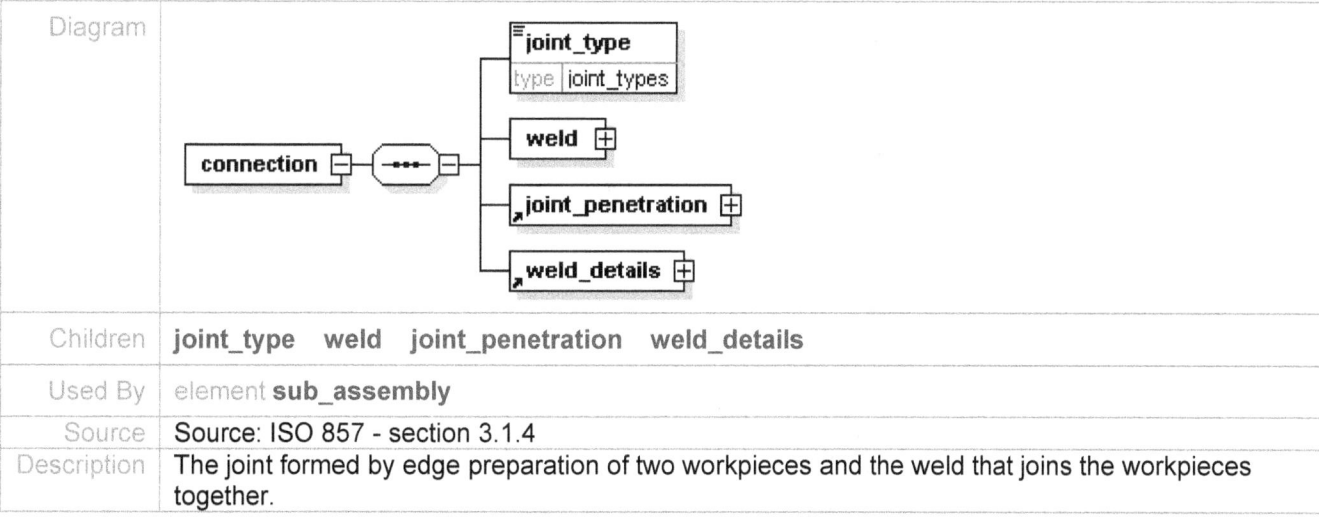
Children	joint_type weld joint_penetration weld_details
Used By	element **sub_assembly**
Source	Source: ISO 857 - section 3.1.4
Description	The joint formed by edge preparation of two workpieces and the weld that joins the workpieces together.

1.2.1 element connection/*joint_type*

Type	joint_types
Values	enumeration butt_joint enumeration corner_joint enumeration t_joint

	enumeration	lap_joint
	enumeration	edge_joint
	enumeration	flanged_butt_joint
	enumeration	flanged_corner_joint
	enumeration	flanged_t_joint
	enumeration	flanged_lap_joint
	enumeration	flanged_edge_joint
Source	AWS A3.0, Figure 1 - Joint Types and definition of term.	
Description	"A weld joint classification based the relative orientation of the members being joined."	

1.2.2 element *connection/weld*

Diagram	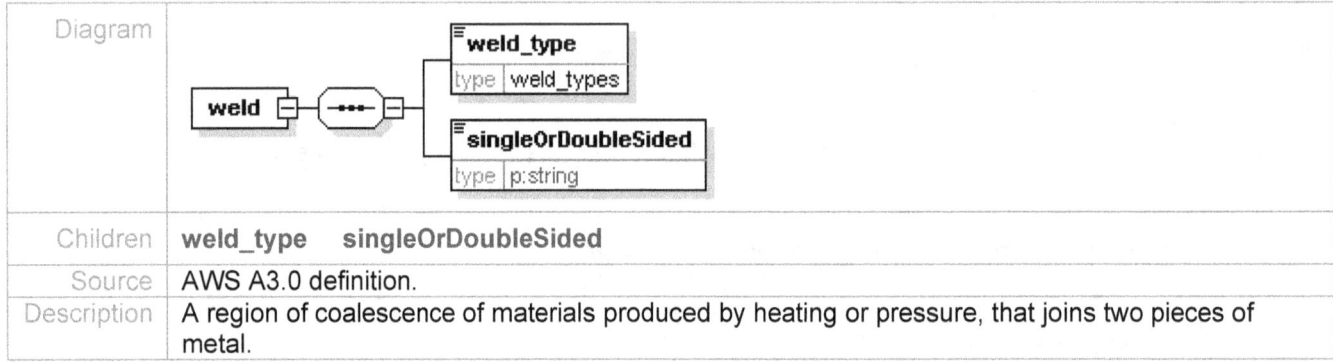
Children	weld_type singleOrDoubleSided
Source	AWS A3.0 definition.
Description	A region of coalescence of materials produced by heating or pressure, that joins two pieces of metal.

1.2.2.1 element connection/weld/**weld_type**

Type	**weld_types**	
Values	enumeration	groove
	enumeration	fillet
	enumeration	plug
	enumeration	slot
	enumeration	spot
	enumeration	seam
	enumeration	flange
	enumeration	stud
	enumeration	surfacing
	enumeration	upset
	enumeration	flash
	enumeration	bevelGroove
	enumeration	flareBevelGroove
	enumeration	doubleBevelGroove
	enumeration	jGroove
	enumeration	singleJGroove
	enumeration	uGroove
	enumeration	singleUGroove
	enumeration	doubleUGroove
	enumeration	vGroove
	enumeration	flareVGroove
	enumeration	singleVGroove
	enumeration	doubleVGroove
	enumeration	squareGroove
	enumeration	edgeFlange
	enumeration	braze

		enumeration	projection
Description		Characterization of the designed weld, by geometry of the objects being joined and of the joint.	

1.2.2.2 element connection/weld/**singleOrDoubleSided**

Type	enumeration of p:string
Values	enumeration single_sided enumeration two_sided
Source	See definiton of "double-welded joint", AWS A3.0, and Figures 8 and 9.
Description	Specification of welding either from only one side or both sides of the joint.

1.3 type **edge_shapes**

Type	enumeration of p:string
Values	enumeration square_edge enumeration single_bevel_edge enumeration double_bevel_edge enumeration single_j_edge enumeration double_j_edge enumeration flanged_edge enumeration round
Source	AWS 3.0, Figure 7 - Edge Shapes
Description	Shape of the edge of a single piece of base material made to form a joint with another piece of base material.

1.4 element **joint_design**

Diagram	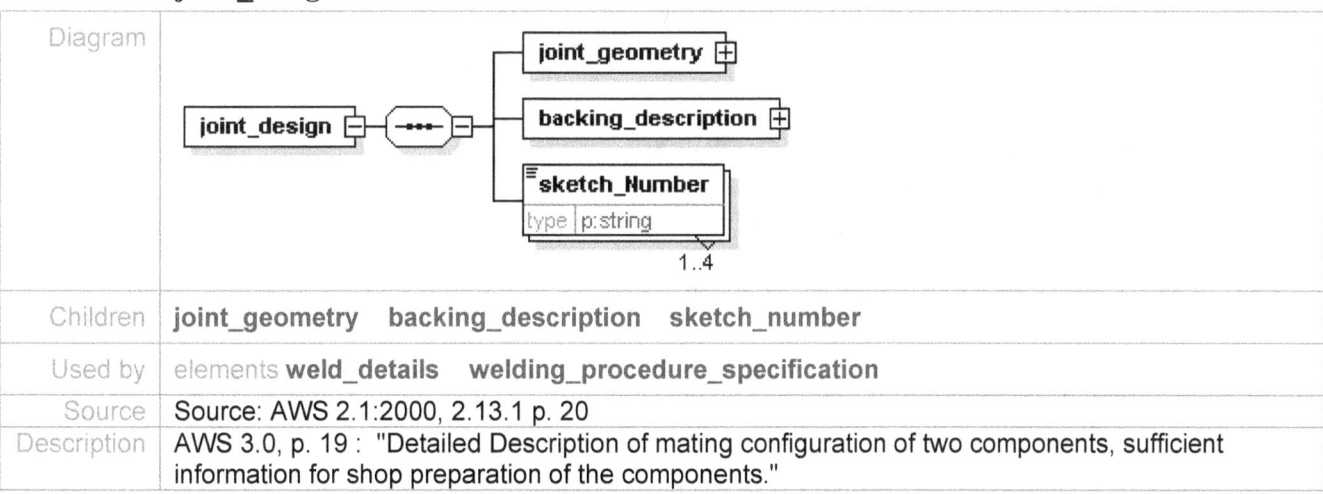
Children	joint_geometry backing_description sketch_number
Used by	elements **weld_details** **welding_procedure_specification**
Source	Source: AWS 2.1:2000, 2.13.1 p. 20
Description	AWS 3.0, p. 19 : "Detailed Description of mating configuration of two components, sufficient information for shop preparation of the components."

1.4.1 element joint_design/**joint_geometry**

Diagram	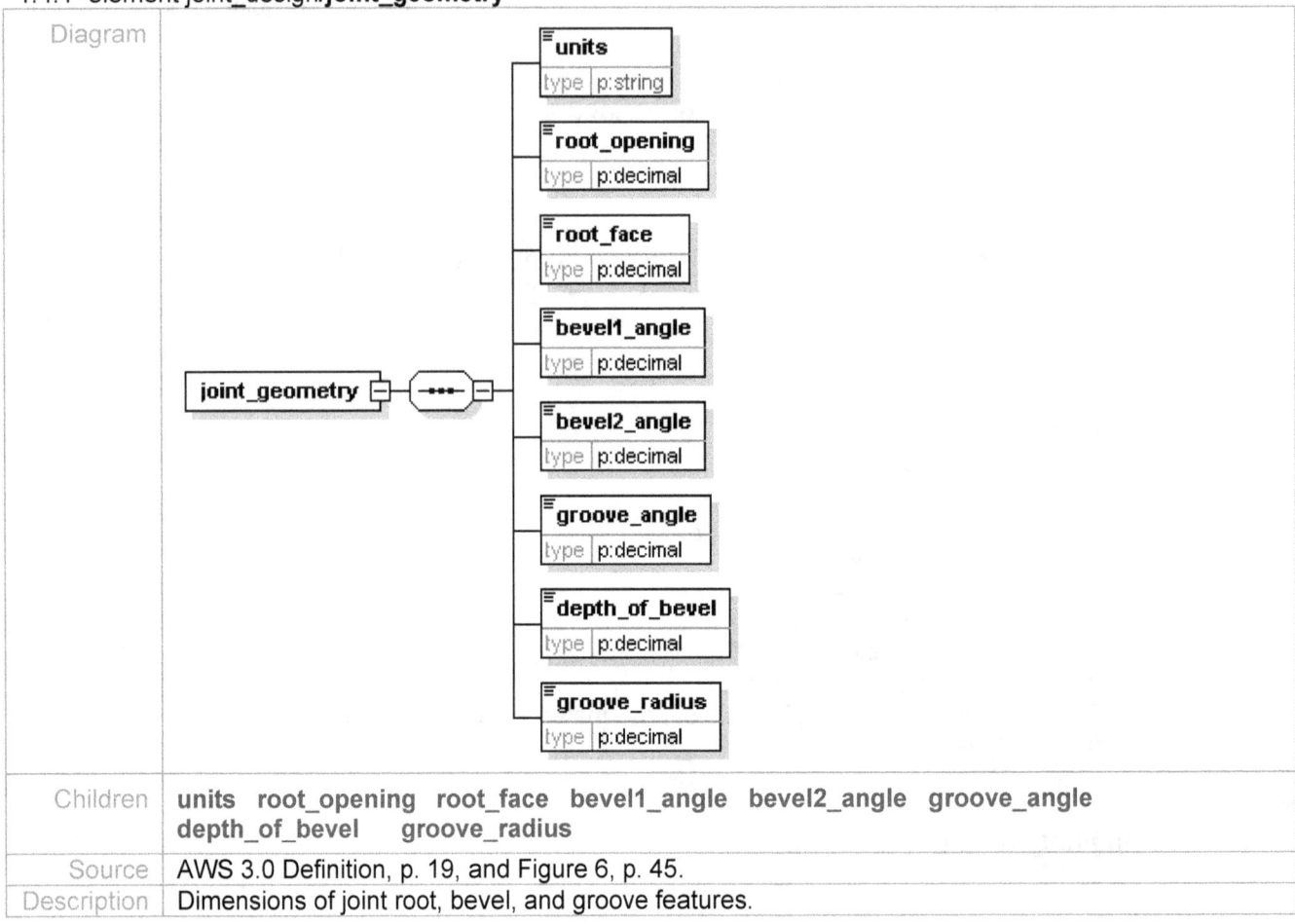
Children	**units root_opening root_face bevel1_angle bevel2_angle groove_angle depth_of_bevel groove_radius**
Source	AWS 3.0 Definition, p. 19, and Figure 6, p. 45.
Description	Dimensions of joint root, bevel, and groove features.

1.4.1.1 element joint_design/joint_geometry/**units**

Type	enumeration of p:string	
Values	enumeration	millimeters
	enumeration	inches
Description	Choice of SI or U.S. Customary units for the linear dimensions.	

1.4.1.2 element joint_design/joint_geometry/**root_opening**

Type	p:decimal
Description	See AWS 3.0, Figure 6.

1.4.1.3 element joint_design/joint_geometry/**root_face**

Type	p:decimal
Source	See AWS A3.0 definition, and Figure 5.
Description	"That portion of the groove face within the joint root".

1.4.1.4 element joint_design/joint_geometry/**bevel1_angle**

Type	p:decimal
Description	See AWS A3.0, Figure 6. Units: degrees.

1.4.1.5 element joint_design/joint_geometry/**bevel2_angle**

Type	p:decimal
Description	See AWS A3.0, Figure 6. Units: degrees.

1.4.1.6 element joint_design/joint_geometry/**groove_angle**

Type	p:decimal
Description	See AWS A3.0, Figure 6. Units: degrees

1.4.1.7 element joint_design/joint_geometry/**depth_of_bevel**

Type	p:decimal
Description	See AWS A3.0, Figure 6. Units: degrees

1.4.1.8 element joint_design/joint_geometry/**groove_radius**

Type	p:decimal
Description	See AWS A3.0, Figure 6. Units: degrees

1.4.2 element joint_design/*backing_description*

Diagram	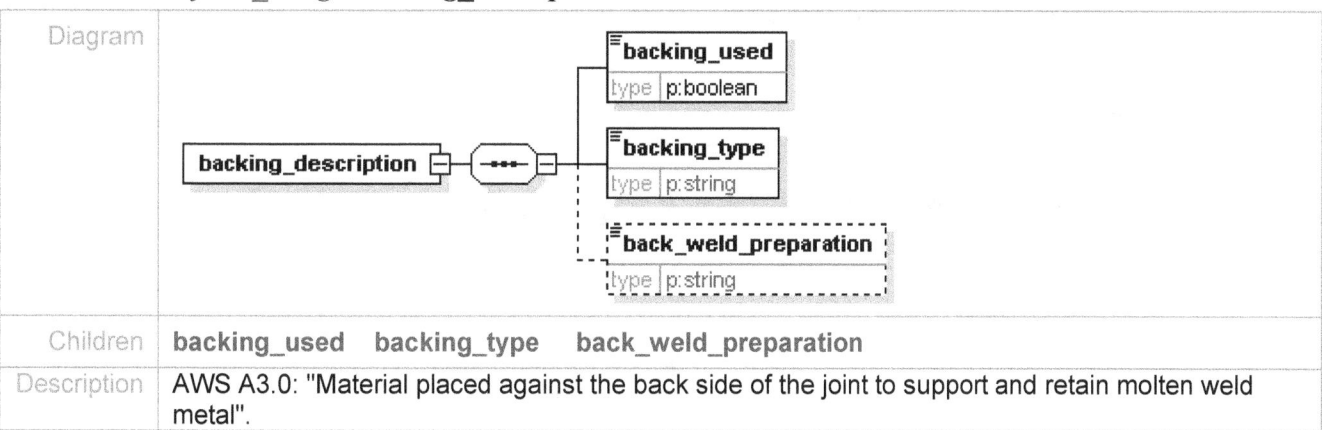
Children	**backing_used backing_type back_weld_preparation**
Description	AWS A3.0: "Material placed against the back side of the joint to support and retain molten weld metal".

1.4.2.1 element joint_design/backing_description/**backing_used**

Type	p:boolean
Description	Boolean stating if backing is required for the weld.

1.4.2.2 element joint_design/backing_description/**backing_type**

Type	enumeration of p:string
Values	enumeration none enumeration strip enumeration strap enumeration ring enumeration split-pipe enumeration weld/sealing run enumeration consumable insert enumeration copper bar enumeration non metallic enumeration retaining shoes enumeration flux backing enumeration backing tape enumeration refractory material enumeration backing_not_alllowed(for PQR)
Source	Source: Welding Handbook, V2, pp. 58, 218, 274. Jefferson's Welding Encyclopedia, p. 46, ISO 9692, ISO 17659.
Description	Most backing Types do not apply to all Types of welding, and should be restricted to the appropriate welding technology (by further refining the definitions, e.g., shoes are typically slag welding only).

1.4.2.3 element joint_design/backing_Description/**back_weld_preparation**

Type	enumeration of p:string
Values	enumeration none enumeration gouging enumeration chipping
Description	AWS A3.0: "Removal of weld metal and base metal from the weld root side of a joint for subsequent welding from that side".

1.4.3 element joint_design/**sketch_number**

Type	p:string
Description	Reference to a sketch of the joint design, the drawing number.

1.5 element **joint_penetration**

Diagram	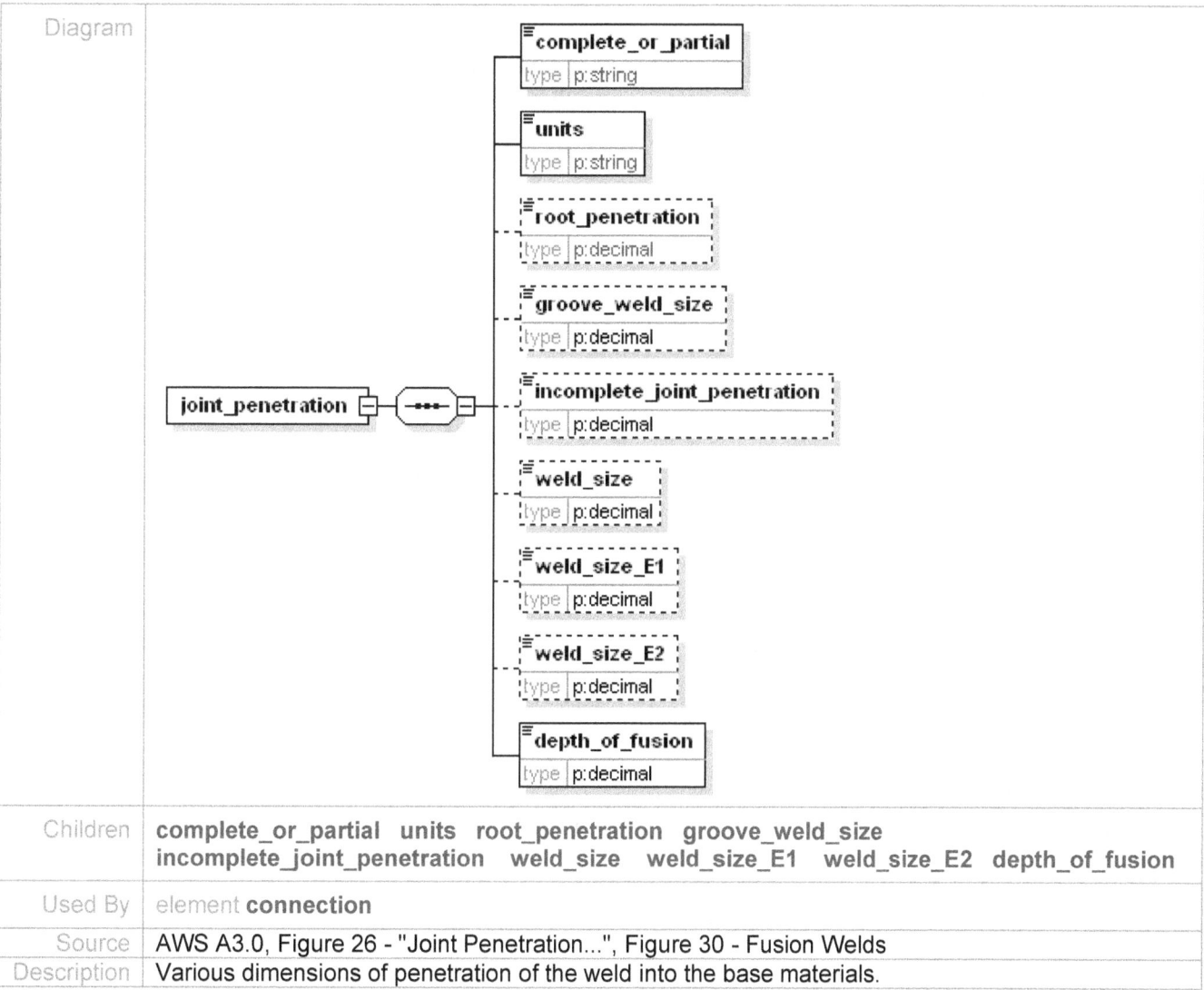
Children	complete_or_partial units root_penetration groove_weld_size incomplete_joint_penetration weld_size weld_size_E1 weld_size_E2 depth_of_fusion
Used By	element **connection**
Source	AWS A3.0, Figure 26 - "Joint Penetration...", Figure 30 - Fusion Welds
Description	Various dimensions of penetration of the weld into the base materials.

1.5.1 element *joint_penetration*/**complete_or_partial**

Type	enumeration of p:string	
Values	enumeration	completePenetration
	enumeration	partialPenetration
Description	The weld design calls for partial or complete penetration.	

1.5.2 element *joint_penetration*/**units**

Type	enumeration of p:string	
Values	enumeration	millimeters
	enumeration	inches
Description	Selection of SI or U.S. Customary units for linear measurements.	

15

1.5.3 element joint_penetration/**root_penetration**

Type	p:decimal
Source	AWS A3.0, definition, and Figure 26 - "Joint Penetration...", Figure 30 - Fusion Welds
Description	"The distance the weld metal extends into the root joint".

1.5.4 element joint_penetration/**groove_weld_size**

Type	p:decimal
Source	AWS A3.0, Figure 26 - "Joint Penetration...", Figure 30 - Fusion Welds
Description	"The joint penetration of a groove weld".

1.5.5 element joint_penetration/**incomplete_joint_penetration**

Type	p:decimal
Description	See AWS A3.0 definition, and Figure 26 - "Joint Penetration...", Figure 30 - Fusion Welds

1.5.6 element joint_penetration/**weld_size**

Type	p:decimal
Description	See AWS A3.0 definition and Figure 26 - "Joint Penetration...", Figure 30 - Fusion Welds

1.5.7 element joint_penetration/**weld_size_E1**

Type	p:decimal
Description	See AWS A3.0 definition and Figure 26 - "Joint Penetration...", Figure 30 - Fusion Welds

1.5.8 element joint_penetration/**weld_size_E2**

Type	p:decimal
Description	See AWS A3.0 definition and Figure 26 - "Joint Penetration...", Figure 30 - Fusion Welds

1.5.9 element joint_penetration/**depth_of_fusion**

Type	p:decimal
Source	AWS 3.0 definition, and Figure 26 - "Joint Penetration...", Figure 30 - Fusion Welds
Description	"The distance that fusion extends into the base metal or the previous bead from the surface melted during welding".

1.6 type **joint_types**

Type	enumeration of p:string
Used By	element **connection/joint_type**

Values	enumeration	butt_joint
	enumeration	corner_joint
	enumeration	t_joint
	enumeration	lap_joint
	enumeration	edge_joint
	enumeration	flanged_butt_joint
	enumeration	flanged_corner_joint
	enumeration	flanged_t_joint
	enumeration	flanged_lap_joint
	enumeration	flanged_edge_joint
Source	AWS A3.0 definition and Figure 2, p. 42	
Description	"A weld joint classification based on relative orientation of members being joined".	

1.7 type linear_dimension_range_type

Diagram	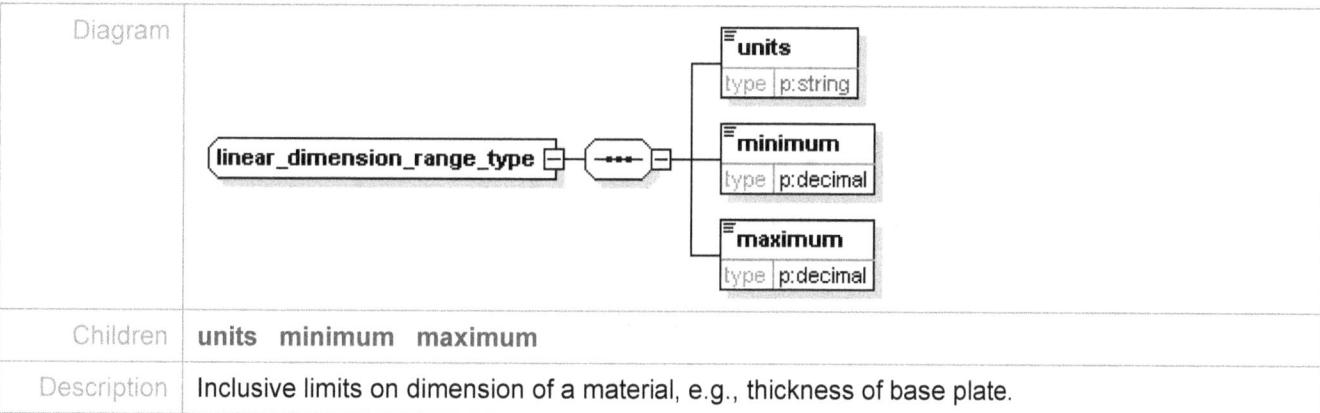
Children	**units minimum maximum**
Description	Inclusive limits on dimension of a material, e.g., thickness of base plate.

1.7.1 element linear_dimension_range_type/*units*

Type	enumeration of p:string
values	enumeration millimeters enumeration inches enumeration gauge
Description	Choice of units to describe a linear dimension.

1.7.2 element linear_dimension_range_type/*minimum*

Type	p:decimal
Description	Inclusive minimum limit.

1.7.3 element linear_dimension_range_type/*maximum*

Type	p:decimal
Description	Inclusive maximum limit.

1.8 type linear_dimension_type_mm

Diagram	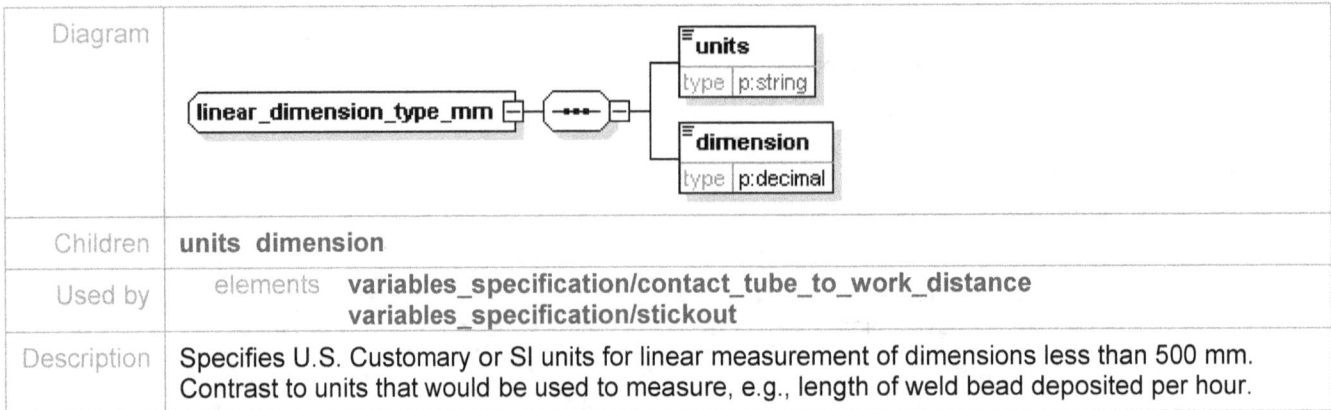
Children	units dimension
Used by	elements **variables_specification/contact_tube_to_work_distance** **variables_specification/stickout**
Description	Specifies U.S. Customary or SI units for linear measurement of dimensions less than 500 mm. Contrast to units that would be used to measure, e.g., length of weld bead deposited per hour.

1.8.1 element linear_dimension_type_mm/units

Type	enumeration of p:string
Values	enumeration millimeters enumeration inches
Description	Choice of SI or U.S. Customary units.

1.8.2 element linear_dimension_type_mm/dimension

Type	p:decimal
Description	The scalar quantity of the measurement.

1.9 element sub_assembly

Diagram	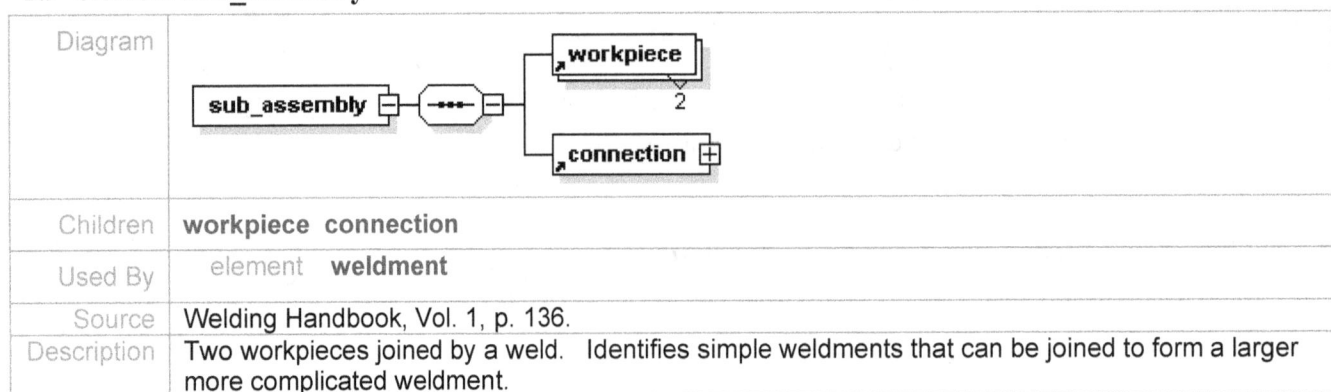
Children	workpiece connection
Used By	element **weldment**
Source	Welding Handbook, Vol. 1, p. 136.
Description	Two workpieces joined by a weld. Identifies simple weldments that can be joined to form a larger more complicated weldment.

1.10 element weld_details

Diagram	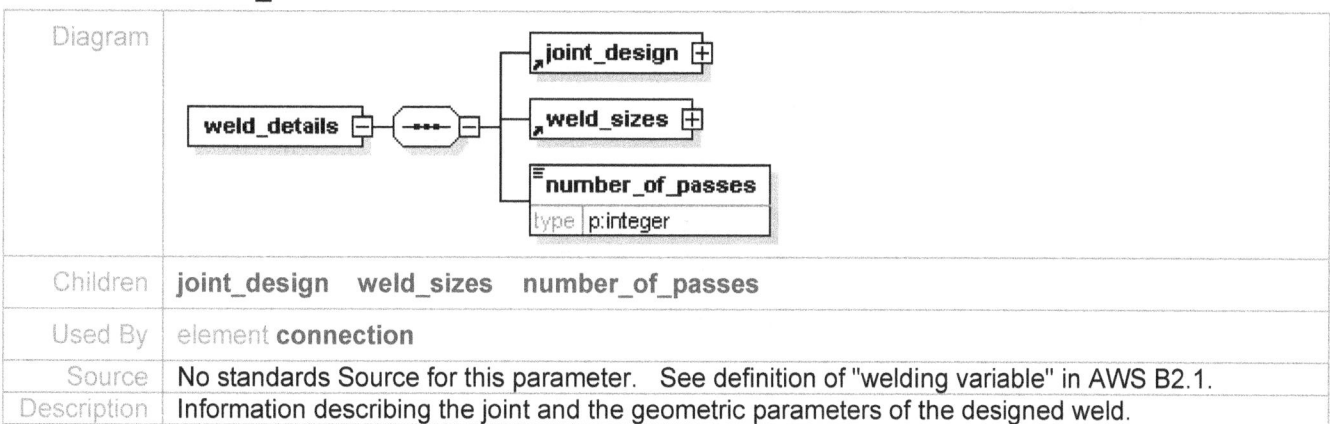
Children	**joint_design weld_sizes number_of_passes**
Used By	element **connection**
Source	No standards Source for this parameter. See definition of "welding variable" in AWS B2.1.
Description	Information describing the joint and the geometric parameters of the designed weld.

1.10.1 element weld_details/*number_of_passes*

Type	p:integer
Description	Number of times needed to deposit weld metal to complete the weld.

1.11 element **weld_sizes**

Diagram	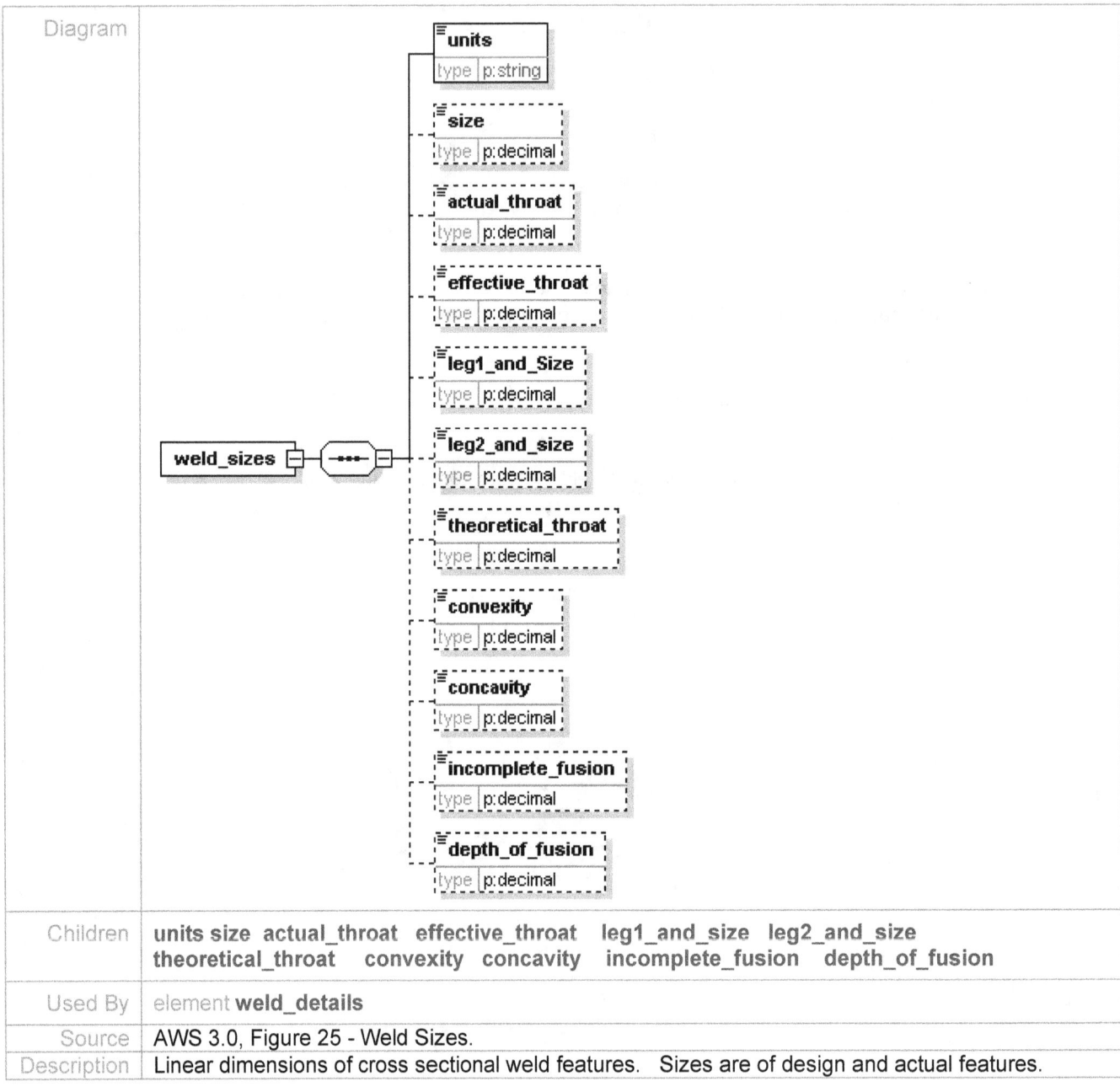
Children	units size actual_throat effective_throat leg1_and_size leg2_and_size theoretical_throat convexity concavity incomplete_fusion depth_of_fusion
Used By	element **weld_details**
Source	AWS 3.0, Figure 25 - Weld Sizes.
Description	Linear dimensions of cross sectional weld features. Sizes are of design and actual features.

1.11.1 element weld_sizes/**units**

Type	enumeration of p:string
Values	enumeration millimeters enumeration inches
Description	Choice of SI or U.S. Customary linear measurement units.

1.11.2 element weld_sizes/*size*

Type	p:decimal
Description	Size of a single weld bead. See AWS 3.0, Figure 25 - Weld Sizes.

1.11.3 element weld_sizes/*actual_throat*

Type	p:decimal
Description	See AWS 3.0, Figure 25 - Weld Sizes.

1.11.4 element weld_sizes/*effective_throat*

Type	p:decimal
Description	See AWS 3.0, Figure 25 - Weld Sizes.

1.11.5 element weld_sizes/*leg1_and_Size*

Type	p:decimal
Description	See AWS 3.0, Figure 25 - Weld Sizes.

1.11.6 element weld_sizes/*leg2_and_size*

Type	p:decimal
Description	See AWS 3.0, Figure 25 - Weld Sizes.

1.11.7 element weld_sizes/*theoretical_throat*

Type	p:decimal
Description	See AWS 3.0, Figure 25 - Weld Sizes.

1.11.8 element weld_sizes/*convexity*

Type	p:decimal
Description	See AWS 3.0, Figure 25 - Weld Sizes.

1.11.9 element weld_sizes/*concavity*

Type	p:decimal
Description	See AWS 3.0, Figure 25 - Weld Sizes.

1.11.10 element weld_sizes/*incomplete_fusion*

Type	p:decimal
Description	See AWS 3.0, Figure 25 - Weld Sizes.

1.11.11 element weld_sizes/*depth_of_fusion*

Type	p:decimal
Description	See AWS 3.0, Figure 25 - Weld Sizes.

1.12 element **weldment**

Children	One or more instances of **sub_assembly**
Source	AWS A3.0
Description	"An assembly whose component parts are joined by welding"

1.13 element **workpiece**

Used By	element **sub_assembly**
Description	This element describes a piece of metal, its material, geometric properties and boundaries, including its edge shape for the welded joint. A workpiece can be a single piece of plate, or a previously assembled weldment. The scope of this data is beyond this welding data dictionary. This element represents an interface with CAD applications.

1.14 type **weld_types**

Type	enumeration of p:string
Used By	element **connection/weld/weld_type**
Values	enumeration groove enumeration fillet enumeration plug enumeration slot enumeration spot enumeration seam enumeration flange enumeration stud enumeration surfacing enumeration upset enumeration flash enumeration bevelGroove enumeration flareBevelGroove enumeration doubleBevelGroove enumeration jGroove enumeration singleJGroove enumeration uGroove enumeration singleUGroove enumeration doubleUGroove

		enumeration	vGroove
		enumeration	flareVGroove
		enumeration	singleVGroove
		enumeration	doubleVGroove
		enumeration	squareGroove
		enumeration	edgeFlange
		enumeration	braze
		enumeration	projection
Source	AWS 3.0 - Figures 8, 9, 15, and AWS A9.1 - Table1		
Description	Classification of the weld by its joint configuration.		

2. Welding Process Data

2.1 type **allied_processes**

Type	enumeration of p:string	
Used By	element **welding_process_type/allied_process**	
Values	enumeration	oxygen_cutting(OC)
	enumeration	arc_cutting(AC)
	enumeration	other_cutting
Source	AWS 3.0, Figure 54b – Master Chart of Allied Processes	
Description	Processes, used by manufacturers that also use welding.	

2.2 type **arc_welding_processes**

Type	enumeration of p:string	
Used By	element **welding_process_type/arc_welding_process**	
Values	enumeration	atomicHydrogenWelding(AHW)
	enumeration	bareMetalArcWelding(BMAW)
	enumeration	carbonArcWelding(CAW)
	enumeration	carbonArcWeldingGas(CAW-G)
	enumeration	carbonArcWeldingShielded(CAW-S)
	enumeration	electrogasWelding(EGW)
	enumeration	electroSlagWelding(ESW)
	enumeration	gasMetalArcWelding(GMAW)
	enumeration	gasTungstenArcWelding(GTAW)
	enumeration	plasmaArcWelding(PAW)
	enumeration	shieldedMetalArcWelding(SMAW)
	enumeration	studArcWelding(SW)
	enumeration	submergedArcWelding(SAW)
	enumeration	submergedArcWeldingSeries(SAW-S)
Description	See AWS 3.0, Figure 54A - Master Chart of Welding and Allied Processes	

2.3 type **brazing_processes**

Type	enumeration of p:string	
Used By	element **welding_process_type/brazing_process**	
Values	enumeration	block_brazing(BB)
	enumeration	diffusion_brazing(CAB)

	enumeration	dip_brazing(DB)
	enumeration	exothermic_brazing(EXB)
	enumeration	flow_brazing(FLOW)
	enumeration	furnace_brazing(FB)
	enumeration	induction_brazing(IB)
	enumeration	infrared_brazing(IRB)
	enumeration	resistance_brazing(RB)
	enumeration	torch_brazing(TB)
	enumeration	twin_carbon_arc_brazing(TCAB)
Description	See AWS 3.0, Figure 54A - Master Chart of Welding and Allied Processes	

2.4 element **electrical_specification**

Diagram	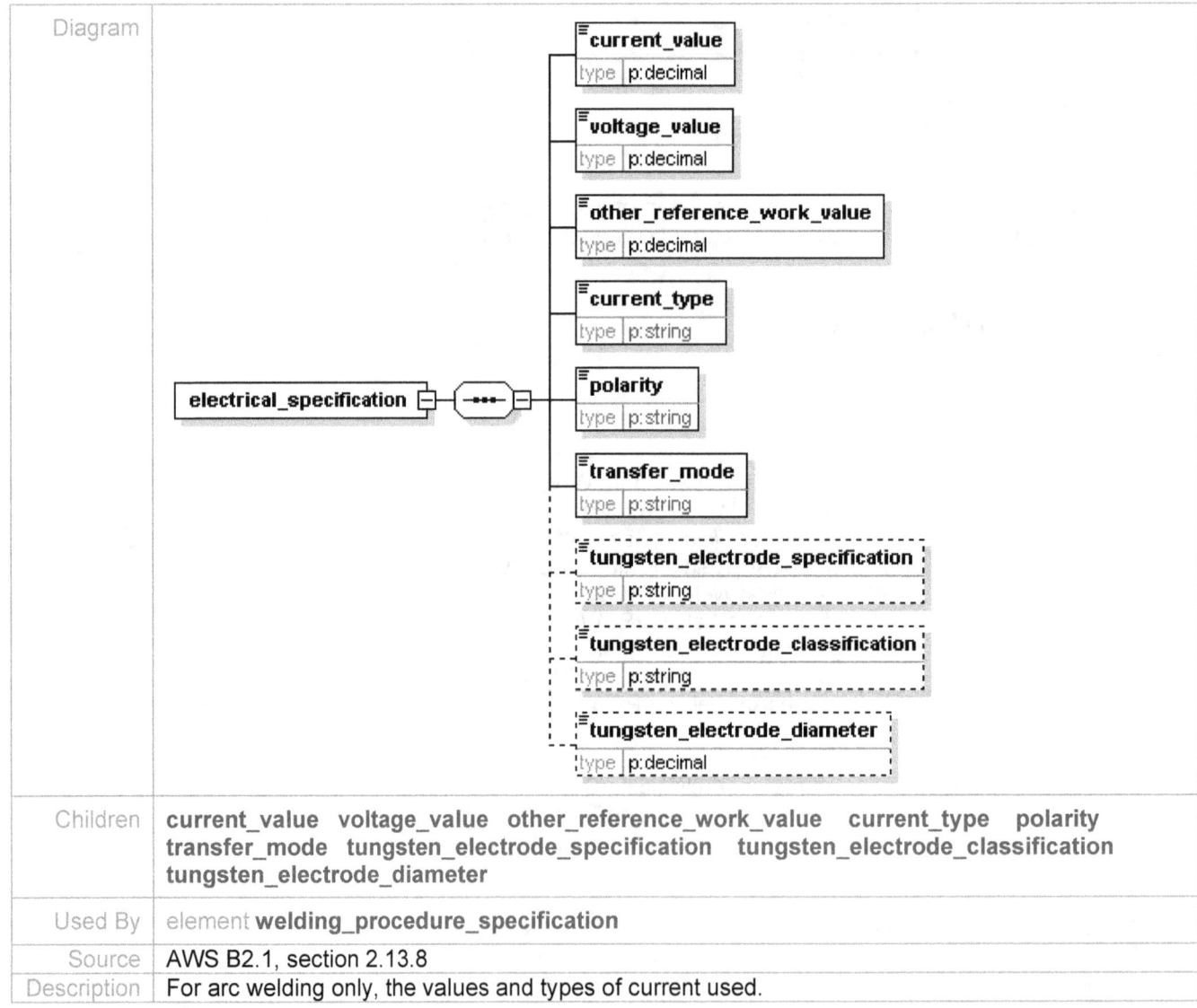
Children	**current_value voltage_value other_reference_work_value current_type polarity transfer_mode tungsten_electrode_specification tungsten_electrode_classification tungsten_electrode_diameter**
Used By	element **welding_procedure_specification**
Source	AWS B2.1, section 2.13.8
Description	For arc welding only, the values and types of current used.

2.4.1 element electrical_specification/*current_value*

Type	p:decimal
Source	AWS B2.1, section 2.13.8 (1)
Description	Current used for a procedure, or current specified (units are amperes).

2.4.2 element electrical_specification/*voltage_value*

Type	p:decimal
Source	AWS B2.1, section 2.13.8 (2)
Description	The voltage to be maintained by the power source, measured between the electrode and the workpiece. Units are volts.

2.4.3 element electrical_specification/*other_reference_work_value*

Type	p:decimal
Description	Commanded variable level for use by various control modes, especially synergic. Units could be WFS, with current levels determined by the intelligent power source.

2.4.4 element electrical_specification/*current_type*

Type	enumeration of p:string
Values	enumeration constant_current enumeration constant_voltage enumeration pulsed
Source	AWS B2.1, section 2.13.8 (1)
Description	The flow of current implemented by the power source.

2.4.5 element electrical_specification/*polarity*

Type	enumeration of p:string
Values	enumeration DCEP enumeration DCEN enumeration AC enumeration PULSED
Source	AWS B2.1, section 2.13.8 (1)
Description	The electrical polarity between the electrode and the workpiece, in direct current arc welding.

2.4.6 element electrical_specification/*transfer_mode*

Type	enumeration of p:string
Values	enumeration spray enumeration globular enumeration short circuit enumeration not applicable
Source	AWS B2.1, section 2.13.8 (5)

Description	The method for causing molten weld metal to leave the consumable electrode and enter the weld pool.

2.4.7 element electrical_specification/*tungsten_electrode_specification*

Type	p:string
Source	AWS A9.1, section 5.1.8, and AWS A5.12.
Description	Description of the electrode's chemical composition, and length and diameter.

2.4.8 element electrical_specification/*tungsten_electrode_classification*

Type	p:string
Source	AWS A9.1, section 5.1.8, and AWS A5.12.
Description	The name of a formal AWS or ISO class, determined by the chemical composition.

2.4.9 element electrical_specification/*tungsten_electrode_diameter*

Type	linear_dimension_type_mm
Source	AWS A9.1, section 5.1.8
Description	Diameter of the electrode.

2.5 element **filler_metal**

Diagram	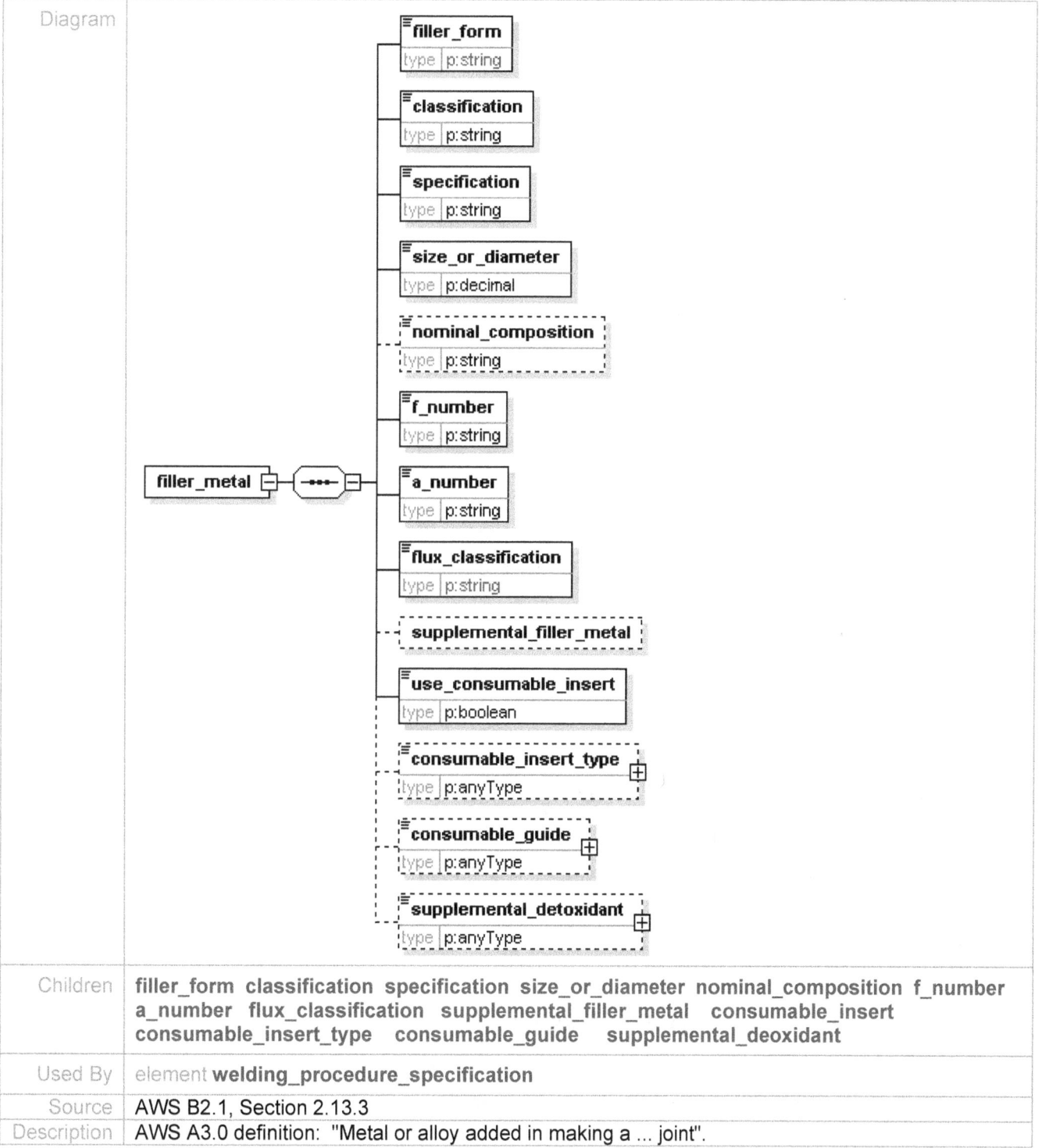
Children	filler_form classification specification size_or_diameter nominal_composition f_number a_number flux_classification supplemental_filler_metal consumable_insert consumable_insert_type consumable_guide supplemental_deoxidant
Used By	element **welding_procedure_specification**
Source	AWS B2.1, Section 2.13.3
Description	AWS A3.0 definition: "Metal or alloy added in making a ... joint".

2.5.1 element filler_metal/*filler_form*

Type	enumeration of p:string
Values	enumeration wire enumeration wire_coiled enumeration wire_spooled enumeration round_wire enumeration rectangular_wire enumeration strip_coiled enumeration strip_spooled enumeration round_wire enumeration rod enumeration rod_rectangular enumeration rod_round enumeration powder_and_paste enumeration foil enumeration sheet
Source	AWS
Description	Physical shape of the filler metal and of its bulk configuration.

2.5.2 element filler_metal/*classification*

Type	p:string
Description	A formal AWS or ISO designation assigned according to the chemical composition.

2.5.3 element filler_metal/*specification*

Type	p:string
Description	A designation including composition and physical properties of the filler metal.

2.5.4 element filler_metal/*size_or_diameter*

Type	linear_dimension_type_mm
Description	Diameter of round metal, or dimensions of prismatic shaped metal.

2.5.5 element filler_metal/*nominal_composition*

Type	p:string
Description	Use only if a formal classification is not used.

2.5.6 element filler_metal/*f_number*

Type	p:string
Description	AWS B2.1 definiton: "A designation used to group welding filler metal for procedure and performance qualifications".

2.5.7 element filler_metal/*a_number*

Type	p:string
Description	AWS B2.1 definition: "a designation to classify ferrous weld metal for procedure qualification based upon chemical composition".

2.5.8 element filler_metal/*flux_classification*

Type	enumeration of p:string	
Values	enumeration	none
	enumeration	solid/metal core
	enumeration	flux_cored
	enumeration	powder_cored
Description	The formal classification of the flux. e.g., ANSI/AWS A5.20 describes classes for carbon steel electrodes for flux cored arc welding.	

2.5.9 element filler_metal/*supplemental_filler_metal*

Type	p:string
Description	Specification of a filler metal used in addition to a consumable electrode.

2.5.10 element filler_metal/*use_consumable_insert*

Type	p:boolean
Description	Boolean stating whether a consumable insert is required or specified.

2.5.11 element filler_metal/*consumable_insert_type*

Type	p:string
Description	Specification of the material property and its size.

2.5.12 element filler_metal/*consumable_guide*

Type	p:string
Description	Material used to contain the weld pool that is melted and becomes part of the weld.

2.5.13 element filler_metal/*supplemental_deoxidant*

Type	p:string
Description	Material added to reduce formation of compounds of oxygen during welding.

2.6 element gas_component

Diagram	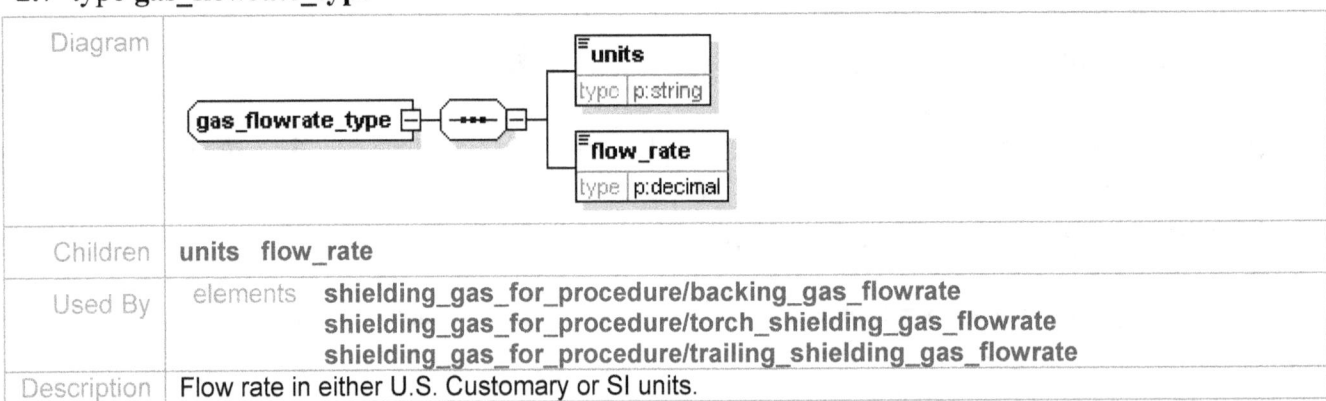
Children	gas_chemical_name gas_percentage
Used By	type **shielding_gas_type**
Source	ANSI/AWS A9.1, section 5.1.7
Description	A single gas element of a mixture and its percentage of the mixture by weight.

2.6.1 element gas_component/*gas_chemical_name*

Type	enumeration of p:string
Values	enumeration argon enumeration carbon dioxide enumeration helium enumeration hydrogen enumeration oxygen
Description	Name of a single element or compound of gas.

2.6.2 element gas_component/*gas_percentage*

Type	p:decimal
Description	Percentage by weight this gas occupies of the total gas mixture.

2.7 type gas_flowrate_type

Diagram	
Children	units flow_rate
Used By	elements shielding_gas_for_procedure/backing_gas_flowrate shielding_gas_for_procedure/torch_shielding_gas_flowrate shielding_gas_for_procedure/trailing_shielding_gas_flowrate
Description	Flow rate in either U.S. Customary or SI units.

2.7.1 element gas_flowrate_type/*units*

Type	enumeration of p:string

Values	enumeration liters per minute
	enumeration cubic feet per hour

2.7.2 element gas_flowrate_type/*flow_rate*

Type	p:decimal

2.8 element **heat_treatment**

Diagram	*(diagram of heat_treatment element showing children: preweld_temperature (temperature_type), preweld_time (p:decimal), max_interpass_temperature (temperature_type), PWHT_minimum_temperature (temperature_type), PWHT_maximum_termperature (temperature_type), PWHT_hold_time (p:decimal))*
Children	preweld_temperature preweld_time max_interpass_temperature PWHT_minimum_temperature PWHT_maximum_termperature PWHT_hold_time
Used By	element **welding_procedure_specification**
Source	AWS B2.1, section 2.13.6. Discussion of preheat in Welding Handbook, V2, pp. 206, 253-255
Description	Description of the heating of a joint (if required) before or after welding, or of maximum or minimum required temperatures to be maintained during welding to improve or maintain desireable properties of the base metal or of the weld.

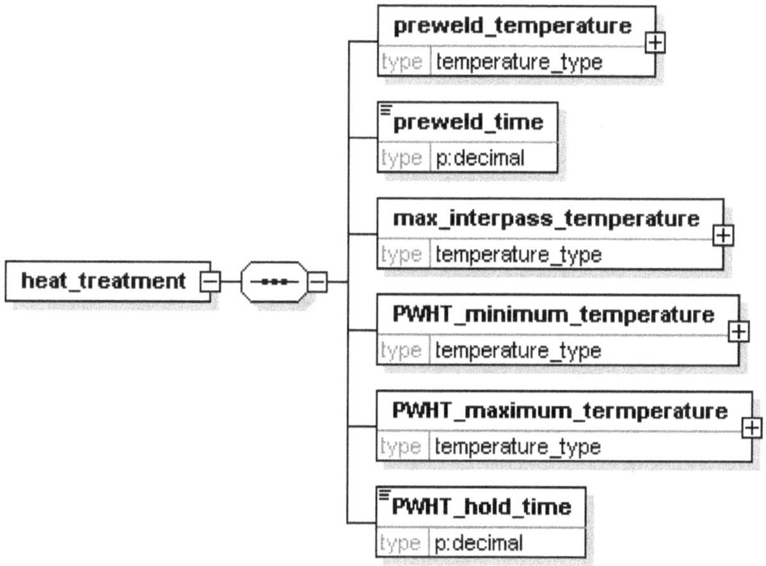

2.8.1 element heat_treatment/*preweld_temperature*

Diagram	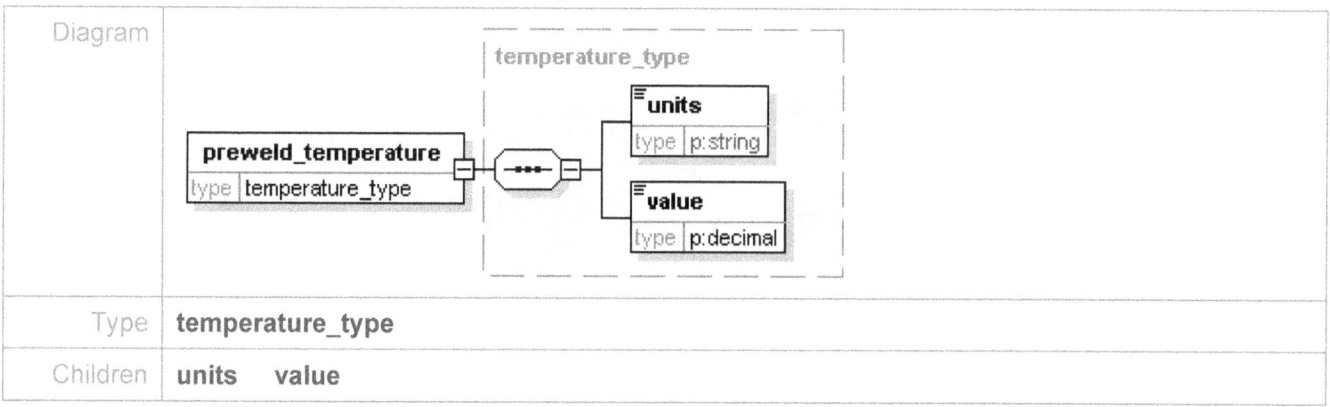
Type	**temperature_type**
Children	**units** **value**

Source	AWB B2.1, section 2.13.5 (1).
Description	Required temperature of base material before welding begins.

2.8.2 element heat_treatment/*preweld_time*

Type	p:decimal
Source	AWS B2.1, section 2.13.5 (3).
Description	Time base materials must be held at the required temperature. Need details on preferred units, e.g., whole minutes, decimal minutes, minutes + seconds.

2.8.3 element heat_treatment/*max_interpass_temperature*

Diagram	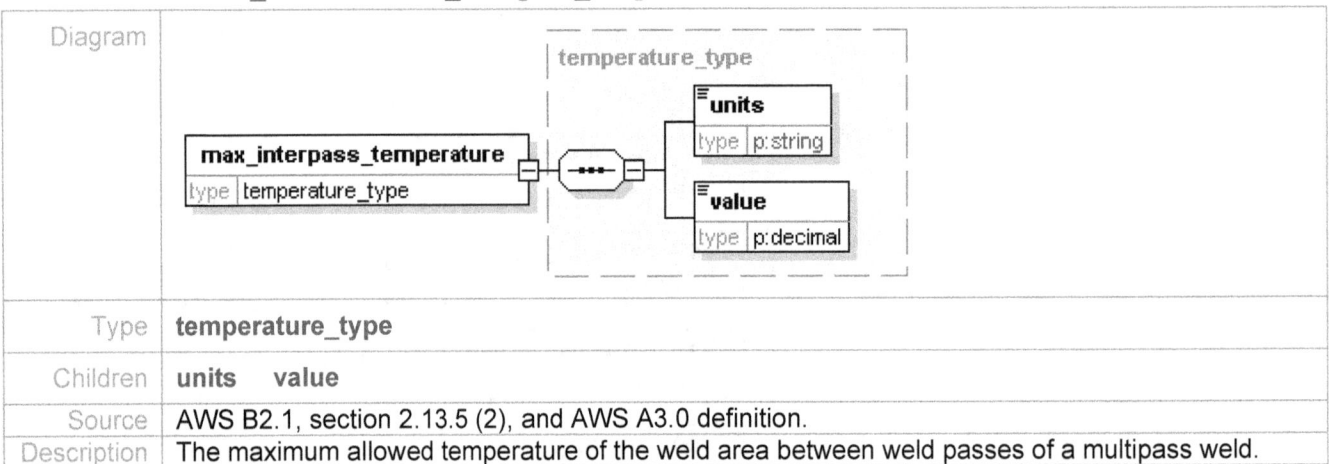
Type	**temperature_type**
Children	**units value**
Source	AWS B2.1, section 2.13.5 (2), and AWS A3.0 definition.
Description	The maximum allowed temperature of the weld area between weld passes of a multipass weld.

2.8.4 element heat_treatment/*PWHT_minimum_temperature*

Diagram	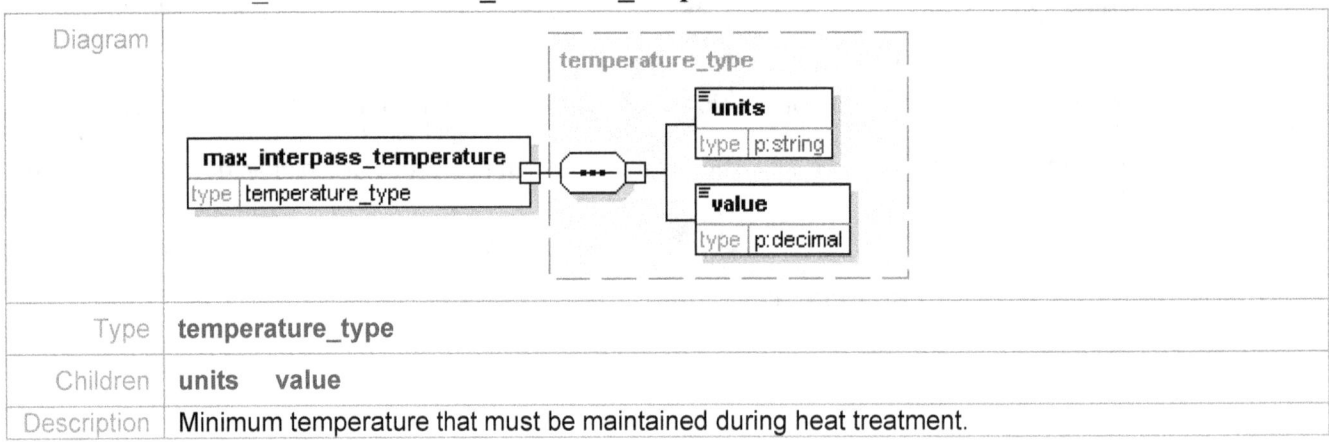
Type	**temperature_type**
Children	**units value**
Description	Minimum temperature that must be maintained during heat treatment.

2.8.5 element heat_treatment/*PWHT_maximum_termperature*

Diagram	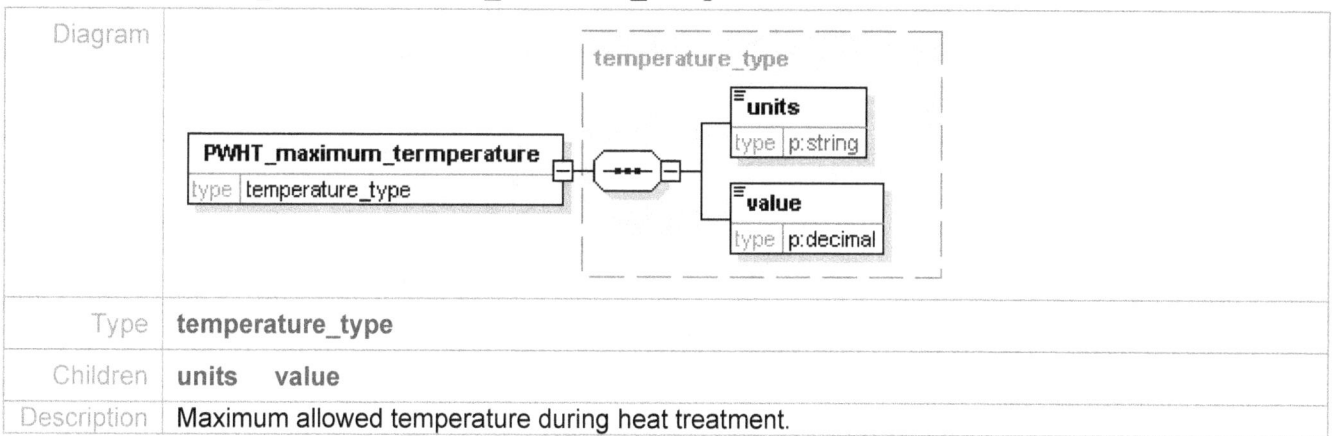
Type	**temperature_type**
Children	**units value**
Description	Maximum allowed temperature during heat treatment.

2.8.6 element heat_treatment/*PWHT_hold_time*

Type	p:decimal
Source	AWS B2.1, section 2.13.6 (1).
Description	Units: minutes

2.9 type **other_welding_processes**

Type	enumeration of p:string
Used By	element **welding_process_type/other_welding_process**
Values	enumeration electron_beam_welding(EBW) enumeration electroslag_welding(ESW) enumeration flow_welding(FLOW) enumeration induction_welding(IW) enumeration laser_beam_welding(LBW) enumeration percussion_welding(PEW) enumeration thermite_welding(TW)
Description	See AWS 3.0, Figure 54A - Master Chart of Welding and Allied Processes

2.10 type **oxyfuel_gas_welding_processes**

Type	enumeration of p:string
Used By	element **welding_process_type/oxyfuel_gas_welding_process**
Values	enumeration air_acetylene_welding(AAW) enumeration oxyacetylene_welding(OAW) enumeration oxyhydrogen_welding(OHW) enumeration pressure_gas_welding(PGW)
Description	See AWS A3.0, Figure 54A - Master Chart of Welding and Allied Processes

2.11 element preheat_and_interpass

Diagram	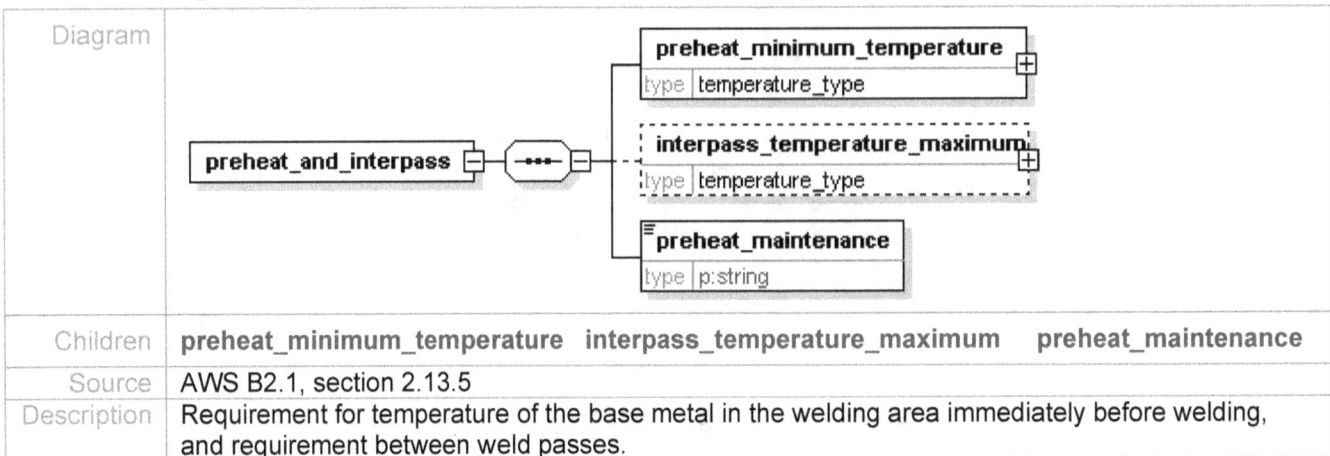
Children	**preheat_minimum_temperature interpass_temperature_maximum preheat_maintenance**
Source	AWS B2.1, section 2.13.5
Description	Requirement for temperature of the base metal in the welding area immediately before welding, and requirement between weld passes.

2.11.1 element preheat_and_interpass/*preheat_minimum_temperature*

Diagram	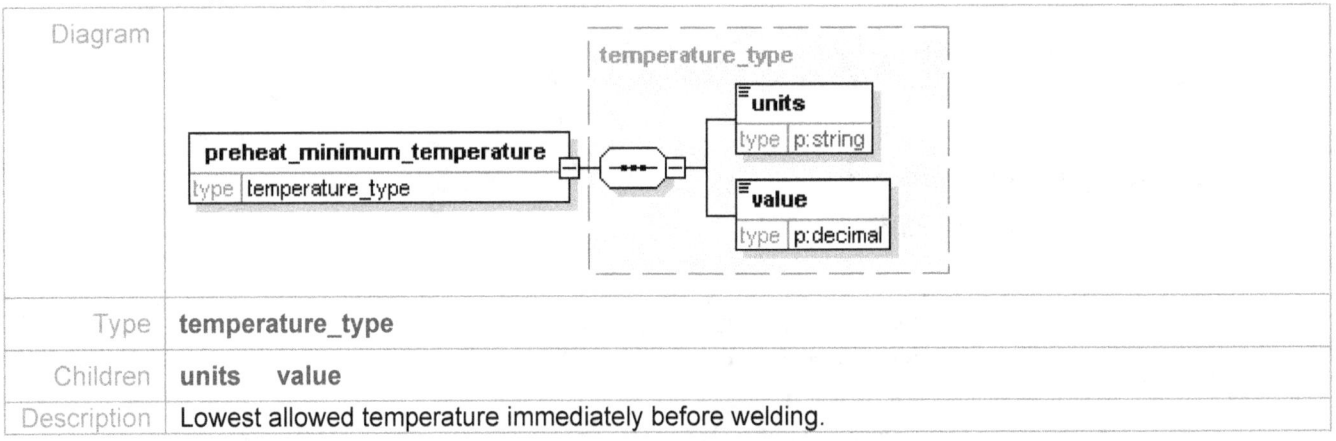
Type	**temperature_type**
Children	**units value**
Description	Lowest allowed temperature immediately before welding.

2.11.2 element preheat_and_interpass/*interpass_temperature_maximum*

Diagram	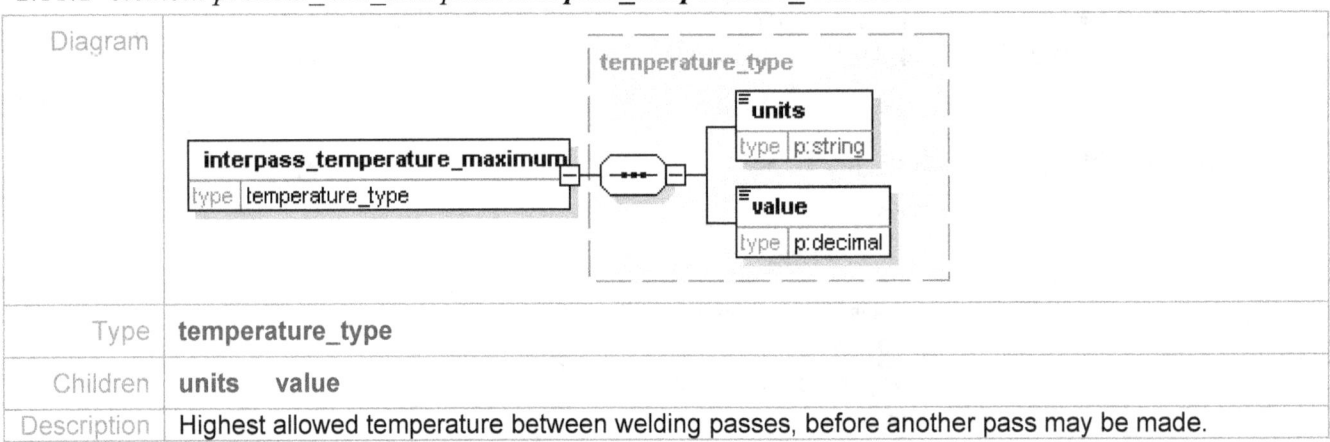
Type	**temperature_type**
Children	**units value**
Description	Highest allowed temperature between welding passes, before another pass may be made.

2.11.3 element preheat_and_interpass/*preheat_maintenance*

Type	enumeration of p:string
Values	enumeration continuous enumeration special heating not required
Description	Method of applying heat to insure weldment stays above minimum allowed temperature.

2.12 type **resistance_welding_processes**

Type	enumeration of p:string
Used By	element **welding_process_type/resistance_welding_process**
Values	enumeration flash_welding(FW) enumeration projection_welding(PW) enumeration resistance_spot_welding(RSW) enumeration resistance_weam_welding(RSEW) enumeration upset_welding(UW)
Description	See AWS 3.0, Figure 54A - Master Chart of Welding and Allied Processes

2.13 element **shielding_gas_for_procedure**

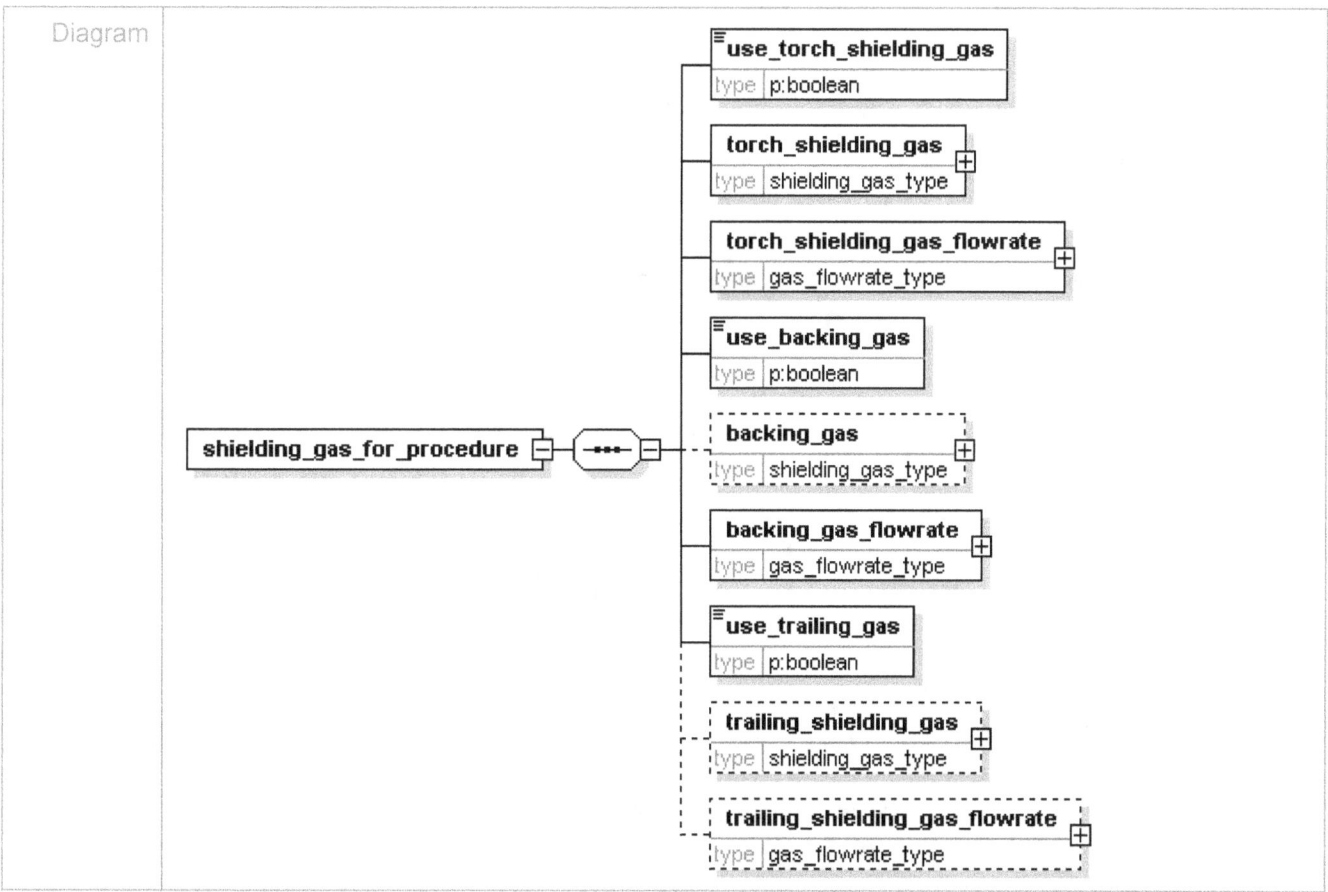

Children	use_torch_shielding_gas torch_shielding_gas torch_shielding_gas_flowrate use_backing_gas backing_gas backing_gas_flowrate use_trailing_gas trailing_shielding_gas trailing_shielding_gas_flowrate
Used By	element **welding_procedure_specification**
Source	AWS B2.1, section 2.13.7
Description	Description of applicable gas composition and flowrates, including torch gas shielding, backing gas, and trailing gas.

2.13.1 element shielding_gas_for_procedure/*use_torch_shielding_gas*

Type	p:boolean
Description	Torch shielding gas is/is not required or specified.

2.13.2 element shielding_gas_for_procedure/*torch_shielding_gas*

Diagram	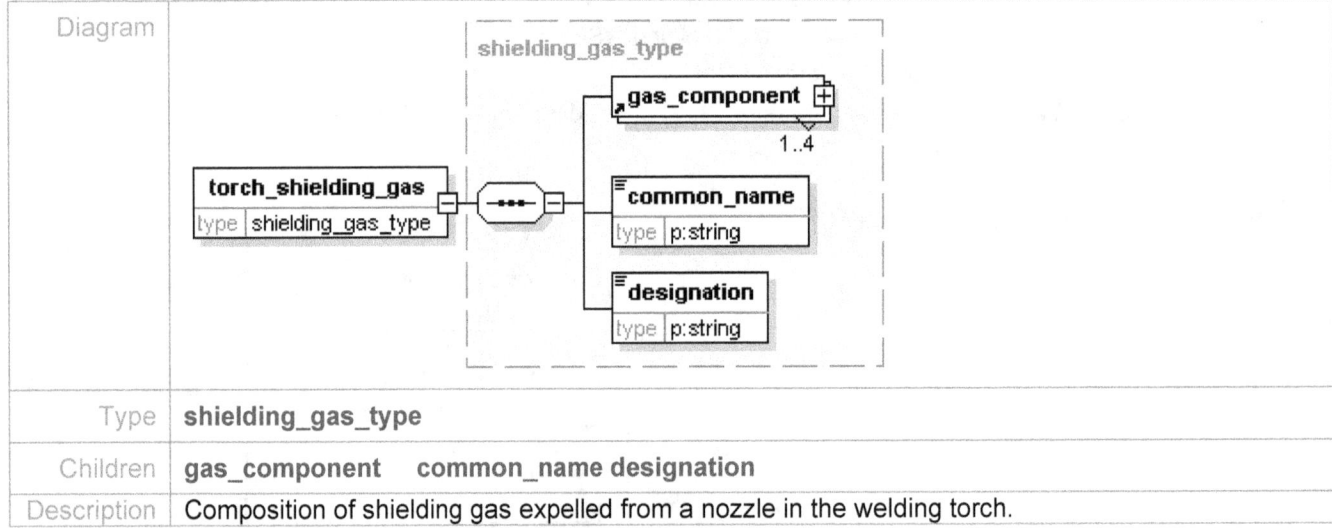
Type	**shielding_gas_type**
Children	**gas_component common_name designation**
Description	Composition of shielding gas expelled from a nozzle in the welding torch.

2.13.3 element shielding_gas_for_procedure/*torch_shielding_gas_flowrate*

Diagram	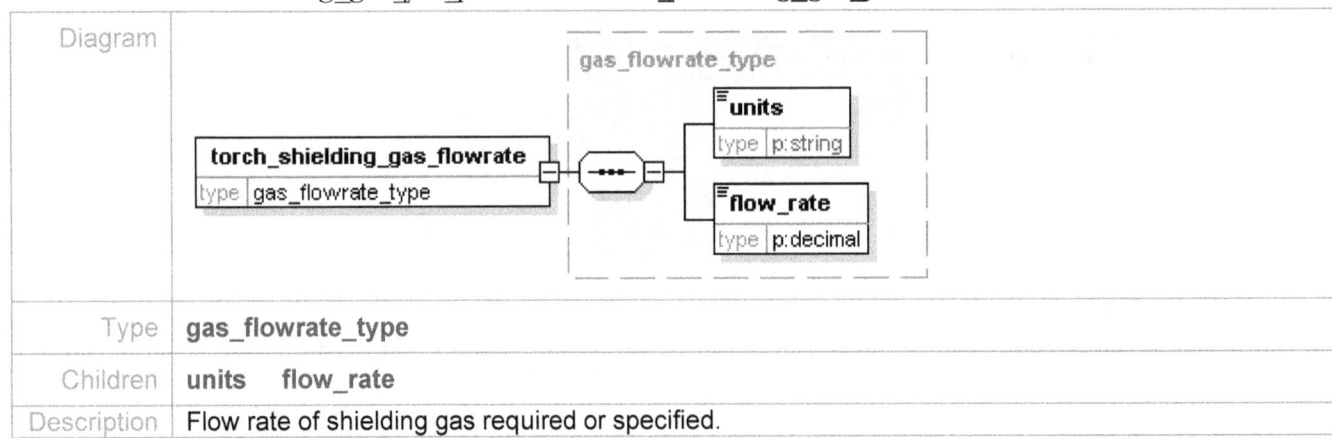
Type	**gas_flowrate_type**
Children	**units flow_rate**
Description	Flow rate of shielding gas required or specified.

2.13.4 element shielding_gas_for_procedure/use_backing_gas

Type	p:boolean
Description	Backing gas is/is not required or specified.

2.13.5 element shielding_gas_for_procedure/backing_gas

Diagram	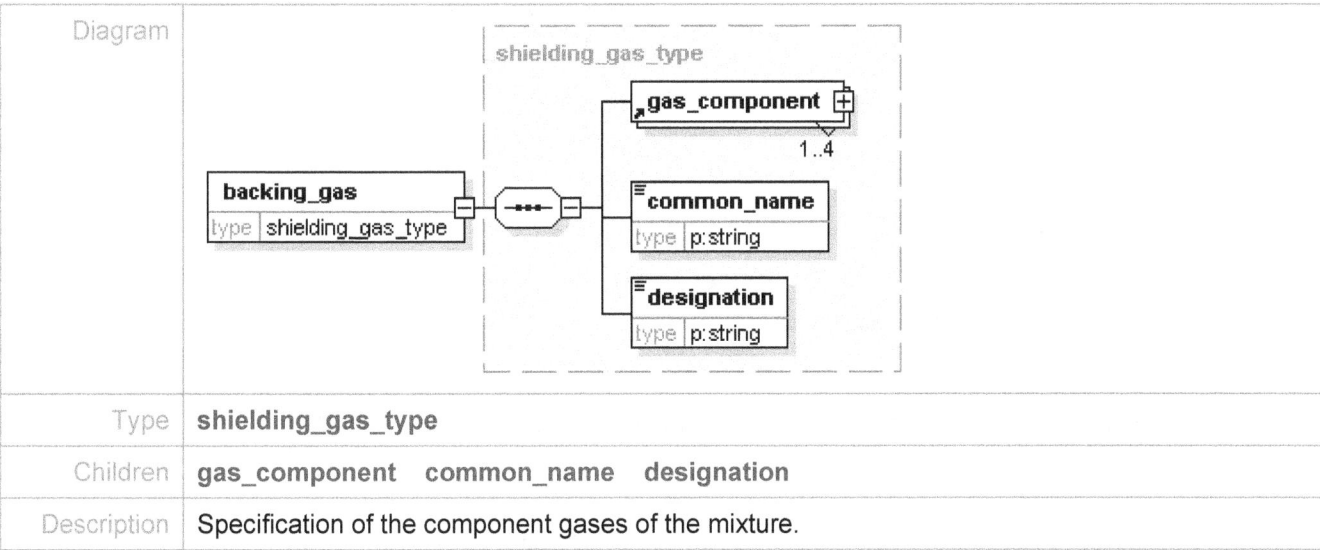
Type	shielding_gas_type
Children	gas_component common_name designation
Description	Specification of the component gases of the mixture.

2.13.6 element shielding_gas_for_procedure/backing_gas_flowrate

Diagram	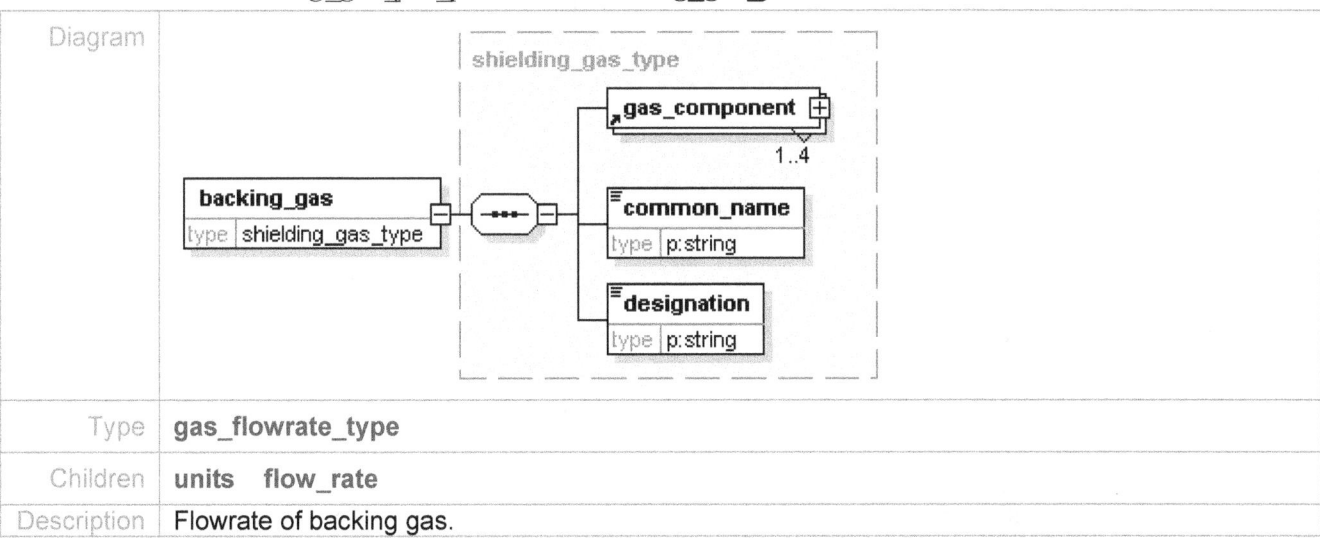
Type	gas_flowrate_type
Children	units flow_rate
Description	Flowrate of backing gas.

2.13.7 element shielding_gas_for_procedure/use_trailing_gas

Type	p:boolean
Description	Trailing shielding gas is/is not required or specified.

2.13.8 element shielding_gas_for_procedure/*trailing_shielding_gas*

Diagram	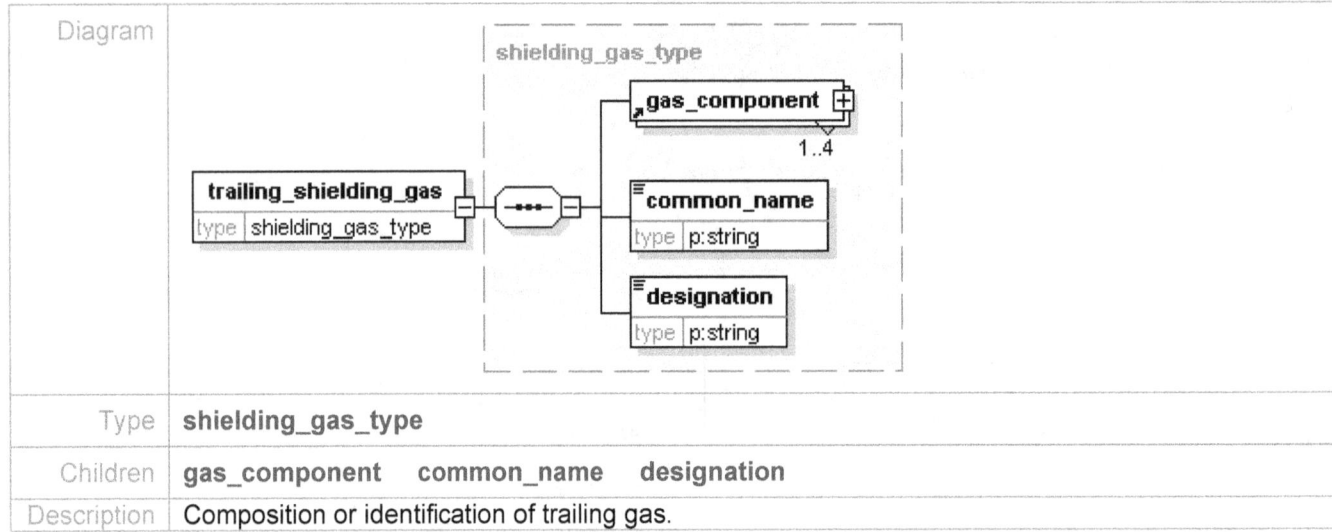
Type	**shielding_gas_type**
Children	**gas_component common_name designation**
Description	Composition or identification of trailing gas.

2.13.9 element shielding_gas_for_procedure/*trailing_shielding_gas_flowrate*

Diagram	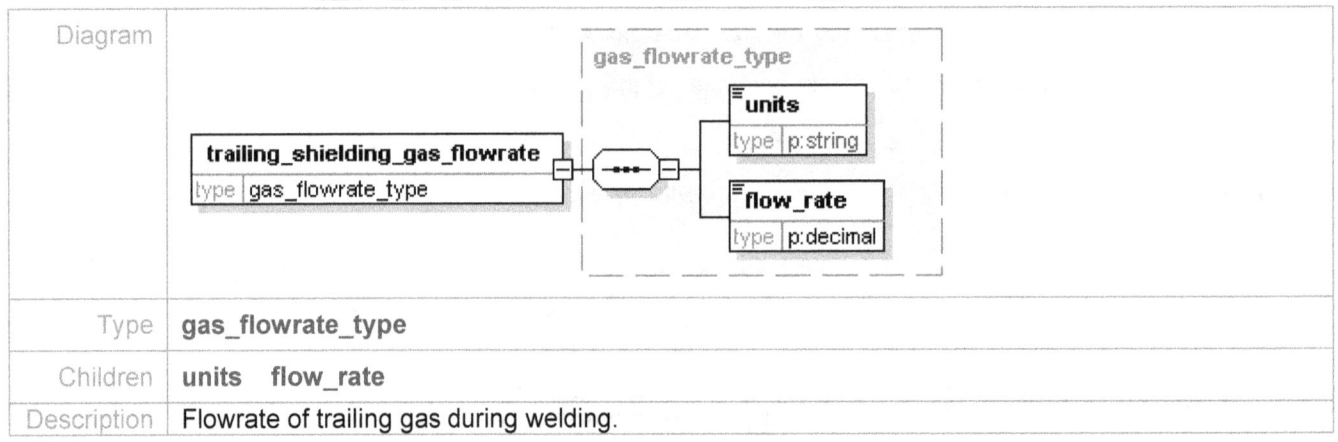
Type	**gas_flowrate_type**
Children	**units flow_rate**
Description	Flowrate of trailing gas during welding.

2.14 type **shielding_gas_type**

Diagram	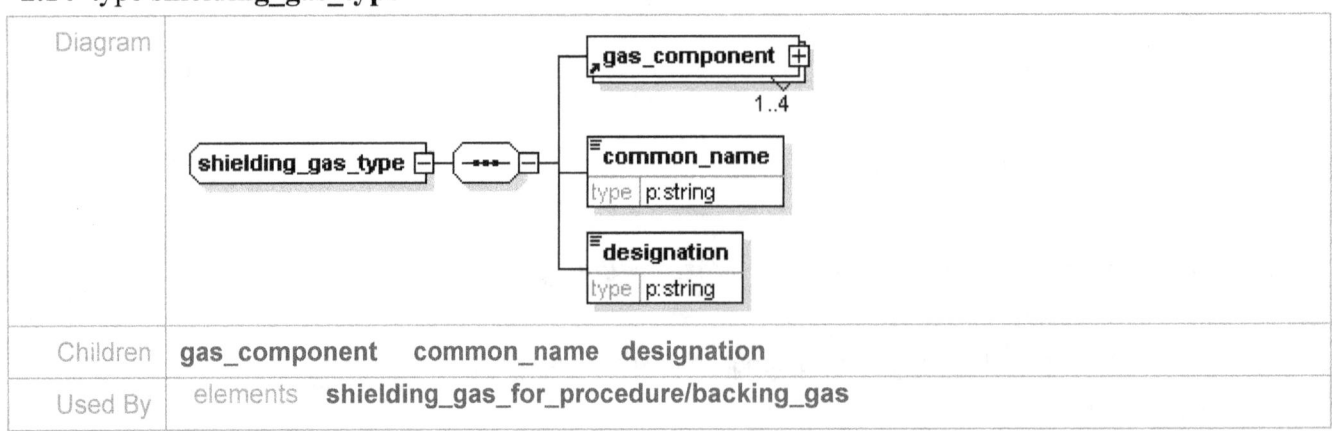
Children	**gas_component common_name designation**
Used By	elements **shielding_gas_for_procedure/backing_gas**

		shielding_gas_for_procedure/torch_shielding_gas shielding_gas_for_procedure/trailing_shielding_gas
Description	Description of a gas or gas mixture used for shielding in arc welding.	

2.14.1 element shielding_gas_type/*common_name*

Type	p:string
Description	Trade name for the gas mixture.

2.14.2 element shielding_gas_type/*designation*

Type	p:string
Description	Specification according to AWS classification by chemical composition of the gas mixture.

2.15 type soldering_processes

Type	p:string	
Used By	element	welding_process_type/soldering_process
Description	See AWS 3.0, Figure 54A - Master Chart of Welding and Allied Processes	

2.16 type solid_state_welding_processes

Type	enumeration of p:string	
Used By	element **welding_process_type/solid_state_welding_process**	
Values	enumeration	coextrusion_welding(CEW)
	enumeration	cold_welding(CW)
	enumeration	diffusion_welding(DFW)
	enumeration	explosion_welding(EXW)
	enumeration	forge_welding(FOW)
	enumeration	friction_welding(FRW)
	enumeration	hot_pressure_welding(HPW)
	enumeration	ultrasonic_welding(USW)
Description	See AWS 3.0, Figure 54A - Master Chart of Welding and Allied Processes	

2.17 type temperature_type

Diagram	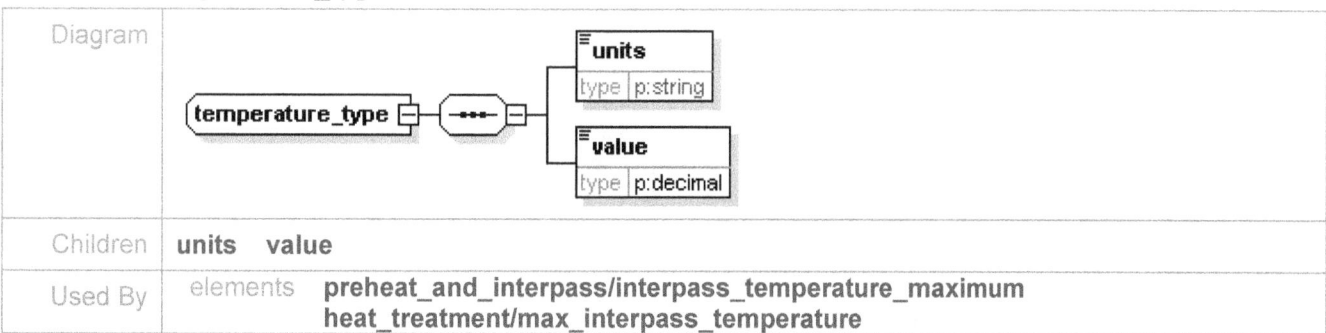
Children	units value
Used By	elements preheat_and_interpass/interpass_temperature_maximum heat_treatment/max_interpass_temperature

		preheat_and_interpass/preheat_minimum_temperature heat_treatment/preweld_temperature heat_treatment/PWHT_maximum_termperature heat_treatment/PWHT_minimum_temperature
Description		Specifies temperature units and the value.

2.17.1 element temperature_type/*units*

Type	enumeration of p:string
Values	enumeration fahrenheit enumeration centigrade
Description	Choice of U.S. Customary or SI units.

2.17.2 element temperature_type/*value*

Type	p:decimal
Description	Numeric value.

2.18 element **variables_specification**

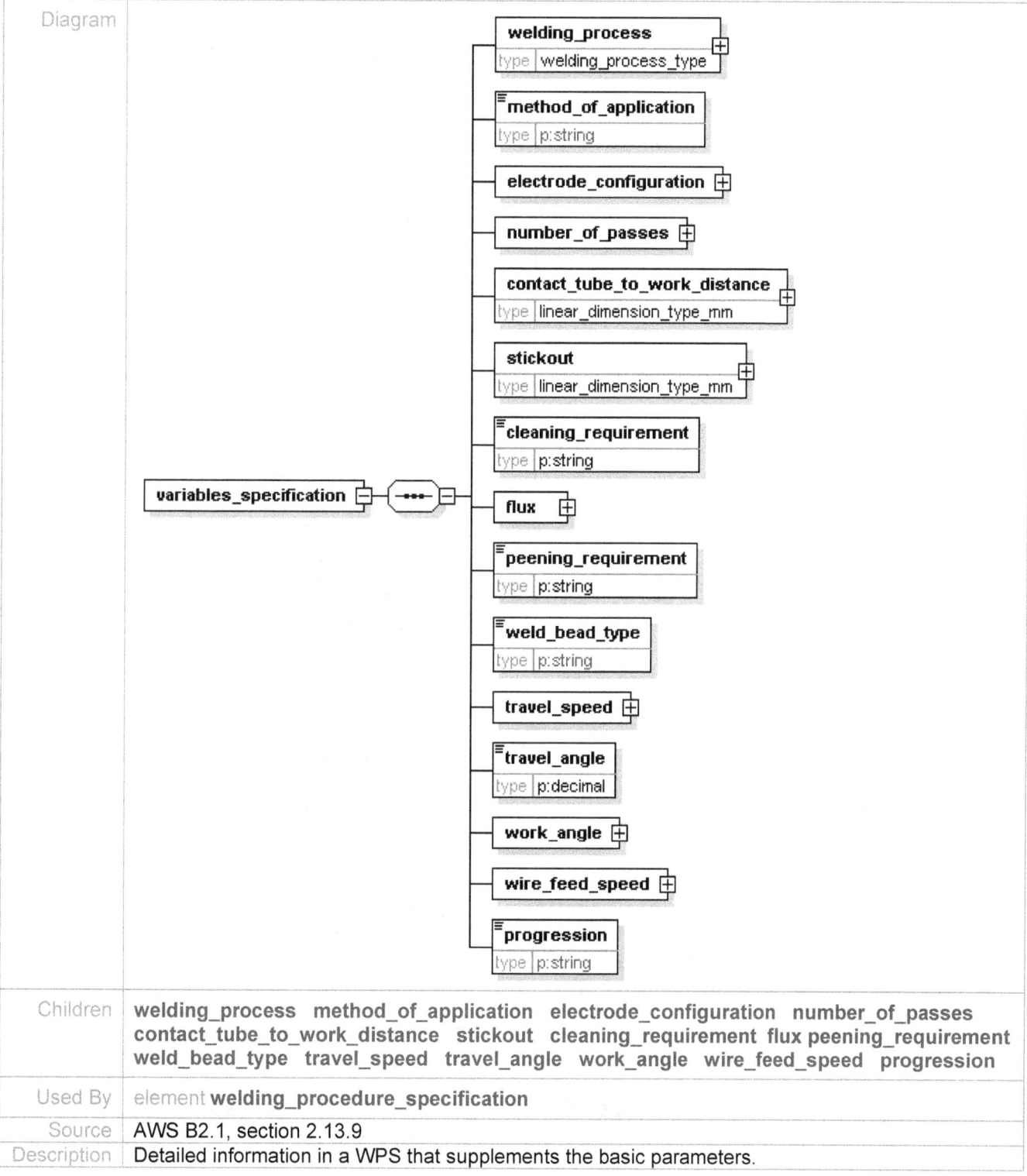

Diagram	
Children	welding_process method_of_application electrode_configuration number_of_passes contact_tube_to_work_distance stickout cleaning_requirement flux peening_requirement weld_bead_type travel_speed travel_angle work_angle wire_feed_speed progression
Used By	element **welding_procedure_specification**
Source	AWS B2.1, section 2.13.9
Description	Detailed information in a WPS that supplements the basic parameters.

2.18.1 element variables_specification/**welding_process**

Diagram	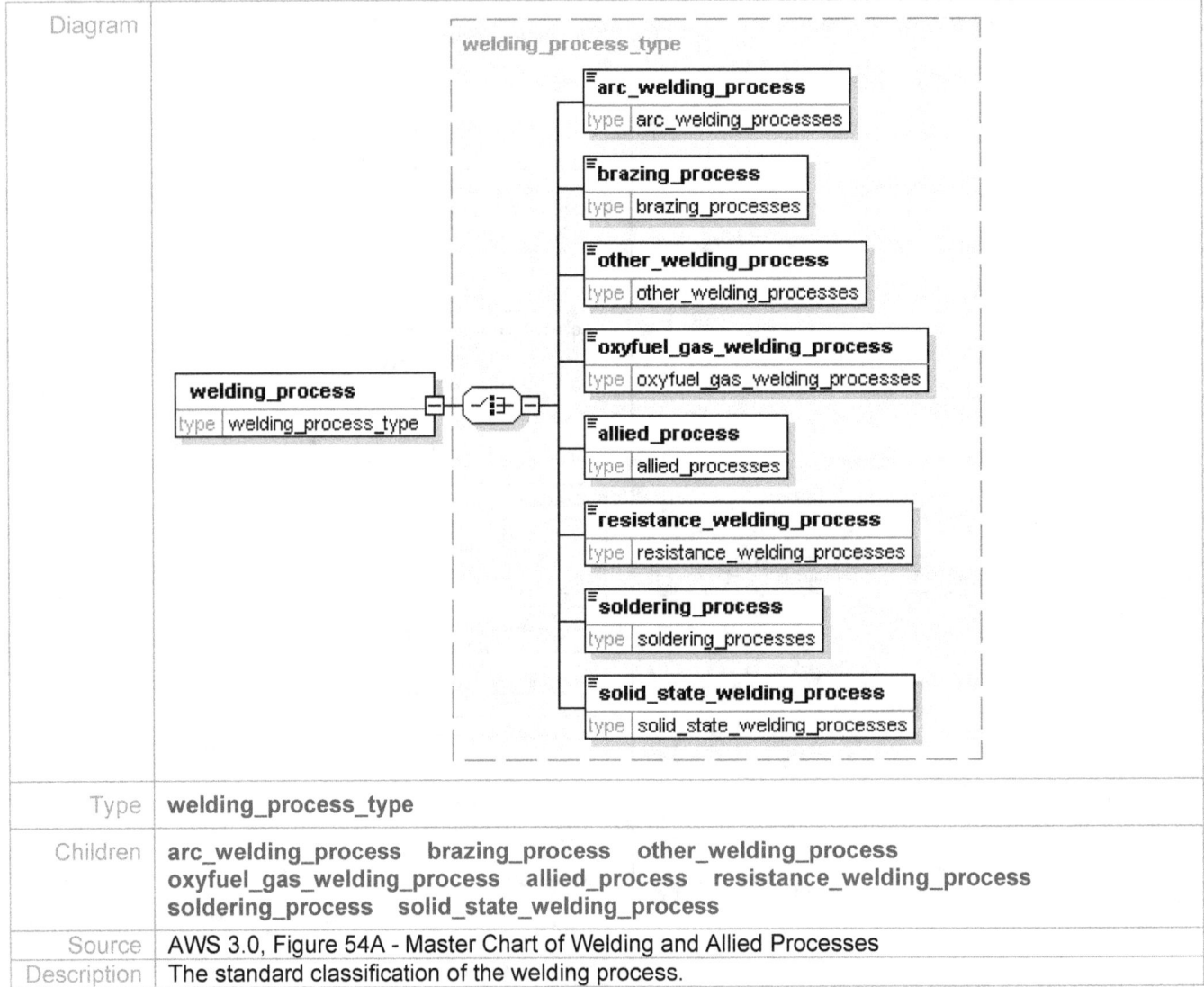
Type	welding_process_type
Children	arc_welding_process brazing_process other_welding_process oxyfuel_gas_welding_process allied_process resistance_welding_process soldering_process solid_state_welding_process
Source	AWS 3.0, Figure 54A - Master Chart of Welding and Allied Processes
Description	The standard classification of the welding process.

2.18.2 element variables_specification/**method_of_application**

Type	enumeration of p:string
Values	enumeration manual enumeration semiautomatic enumeration mechanized enumeration automatic
Source	AWS B2.1, section 2.13.9
Description	The means of manipulating the welding electrode, either by person or machine.

2.18.3 element variables_specification/*electrode_configuration*

Diagram	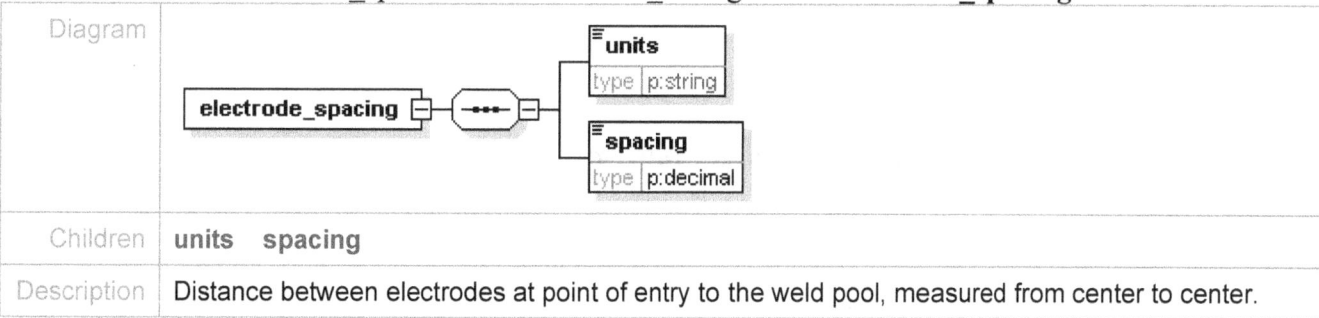
Children	**number_of_electrodes** **electrode_spacing**
Source	AWS B2.1, section 2.13.9(2)
Description	For non-manual welding, use of single or multiple electrodes

2.18.3.1 element variables_specification/electrode_configuration/**number_of_electrodes**

Type	p:integer
Description	Number of electrodes that conduct welding current simultaneously.

2.18.3.2 element variables_specification/electrode_configuration/**electrode_spacing**

Diagram	
Children	**units** **spacing**
Description	Distance between electrodes at point of entry to the weld pool, measured from center to center.

2.18.3.2.1 element variables_specification/electrode_configuration/electrode_spacing/**units**

Type	enumeration of p:string
Values	enumeration millimeters enumeration inches

2.18.3.2.2 element variables_specification/electrode_configuration/electrode_spacing/**spacing**

Type	linear_dimension_type_mm
Description	Distance from center to center between dual electrodes.

2.18.4 element variables_specification/**number_of_passes**

Diagram	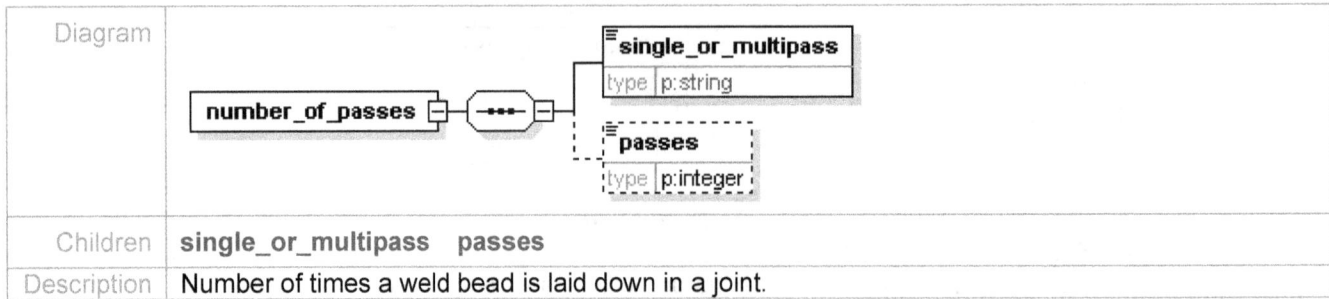
Children	**single_or_multipass passes**
Description	Number of times a weld bead is laid down in a joint.

2.18.4.1 element variables_specification/number_of_passes/**single_or_multipass**

Type	enumeration of p:string
Values	enumeration single_pass enumeration multi_pass

2.18.4.2 element variables_specification/number_of_passes/**passes**

Type	p:integer

2.18.5 element variables_specification/**contact_tube_to_work_distance**

Diagram	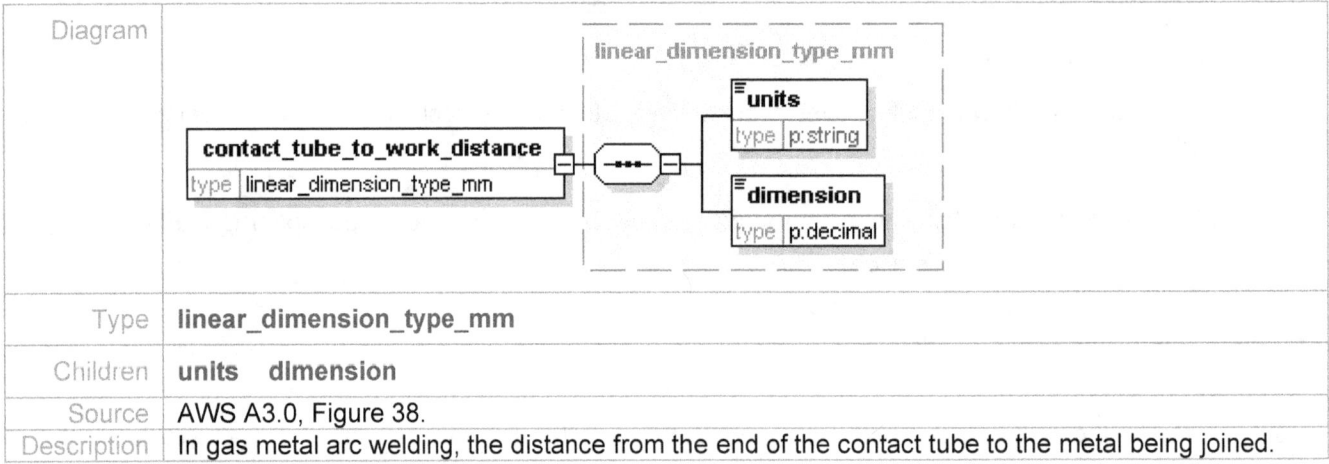
Type	**linear_dimension_type_mm**
Children	**units dimension**
Source	AWS A3.0, Figure 38.
Description	In gas metal arc welding, the distance from the end of the contact tube to the metal being joined.

2.18.6 element variables_specification/*stickout*

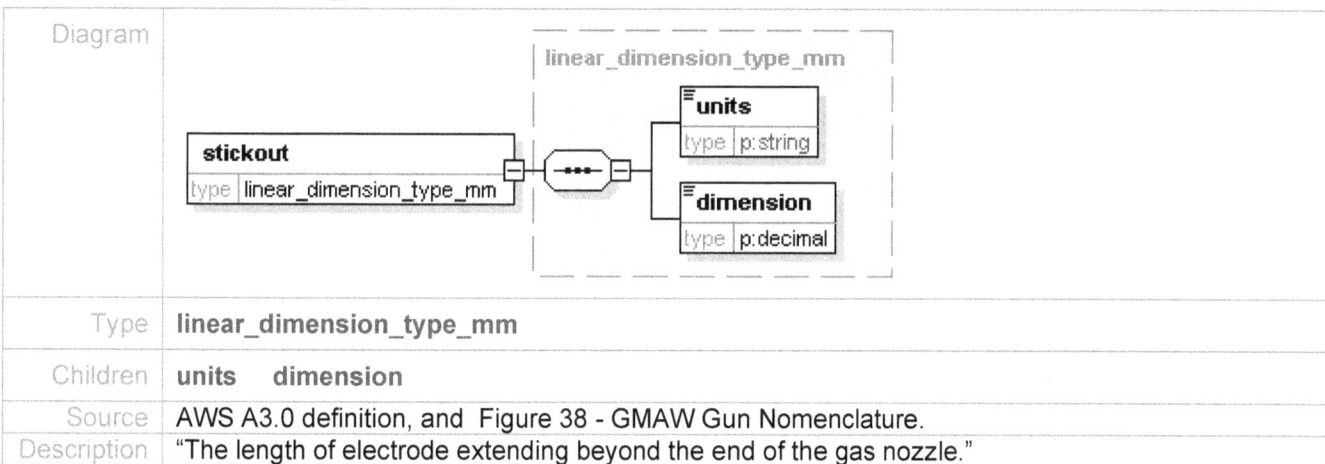

Type	linear_dimension_type_mm
Children	units dimension
Source	AWS A3.0 definition, and Figure 38 - GMAW Gun Nomenclature.
Description	"The length of electrode extending beyond the end of the gas nozzle."

2.18.7 element variables_specification/*cleaning_requirement*

Type	p:string
Source	AWS B2.1, section 2.13.9 (5)
Description	Required method of cleaning before welding, done to ensure weld quality.

2.18.8 element variables_specification/*flux*

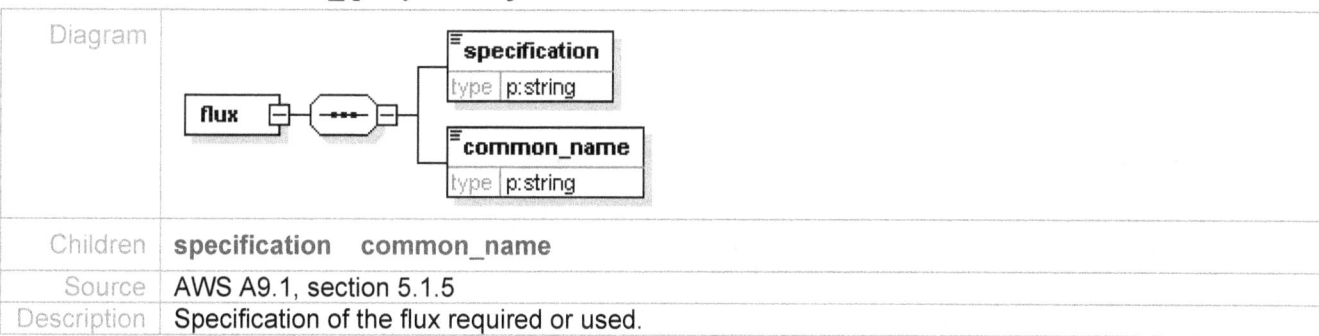

Children	specification common_name
Source	AWS A9.1, section 5.1.5
Description	Specification of the flux required or used.

2.18.8.1 element variables_specification/flux/*specification*

Type	p:string
Description	AWS or ISO designation for the flux composition.

2.18.8.2 element variables_specification/flux/**common_name**

Type	p:string

2.18.9 element variables_specification/**peening_requirement**

Type	p:string
Source	AWS B2.1, section 2.13.9 (6).
Description	Statement of type of hammered working of the completed weld metal needed, if any.

2.18.10 element variables_specification/**weld_bead_type**

Type	enumeration of p:string	
Values	enumeration	stringer bead
	enumeration	weave bead
Source	AWS A3.0 - Figure 22, AWS B2.1, section 2.13.9	
Description	A bead produced by straight line motion or oscillation of the torch.	

2.18.11 element variables_specification/**travel_speed**

Diagram	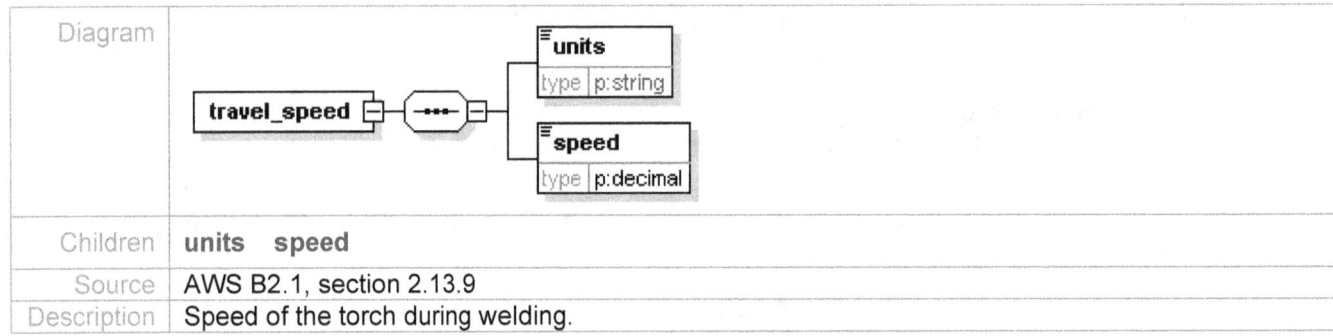
Children	units speed
Source	AWS B2.1, section 2.13.9
Description	Speed of the torch during welding.

2.18.11.1 element variables_specification/travel_speed/**units**

Type	enumeration of p:string	
Values	enumeration	inches per minute
	enumeration	millimeters per minute
Description	Choice of U.S. Customary or SI units.	

2.18.11.2 element variables_specification/travel_speed/**speed**

Type	p:decimal
Description	Linear velocity of the torch in relation to the workpiece.

2.18.12 element variables_specification/**travel_angle**

Type	p:decimal
Source	AWS A9.1, section 5.2.5.9
Description	AWS A3.0 definition: "The angle between the electrode axis and a line perpendicular to the weld axis".

2.18.13 element variables_specification/work_angle

Diagram	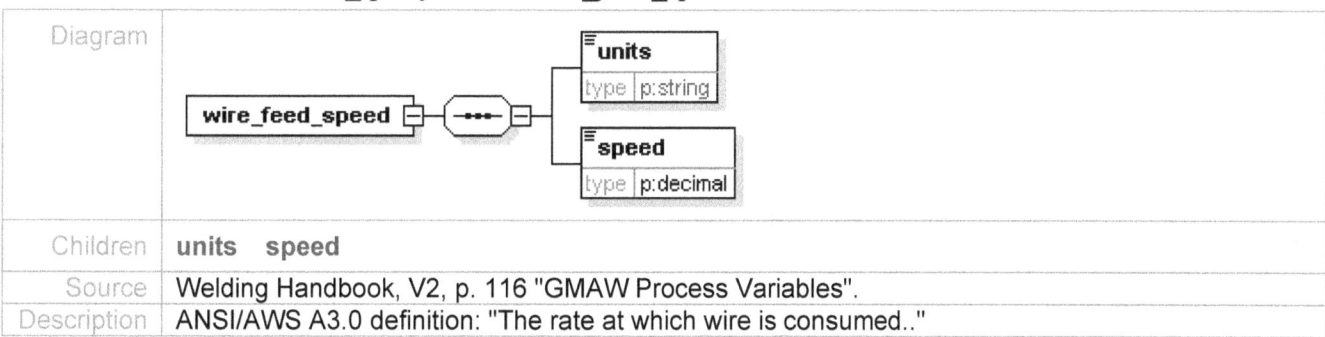
Children	**push_or_drag** **angle**
Source	AWS A9.1, section 5.2.5.9.
Description	AWS A3.0 definition: "The angle between a line perpendicular to the major workpiece surface and a plane determined by the electrode axis and the weld axis"

2.18.13.1 element variables_specification/work_angle/**push_or_drag**

Type	enumeration of p:string
Values	enumeration push
	enumeration drag

2.18.13.2 element variables_specification/work_angle/**angle**

Type	p:decimal

2.18.14 element variables_specification/**wire_feed_speed**

Diagram	
Children	**units** **speed**
Source	Welding Handbook, V2, p. 116 "GMAW Process Variables".
Description	ANSI/AWS A3.0 definition: "The rate at which wire is consumed.."

2.18.14.1 element variables_specification/wire_feed_speed/**units**

Type	enumeration of p:string
Values	enumeration millimeters per second
	enumeration inches per minute

2.18.14.2 element variables_specification/wire_feed_speed/**speed**

Type	p:decimal

*2.18.15 element variables_specification/**progression***

Type	enumeration of p:string
Values	enumeration upward enumeration downward
Source	AWS B2.1, section 2.13.4 (2)
Description	The movement of the torch, or the orientation of the weld joint, in relation to the direction of gravity.

2.19 element **welding_position**

Diagram	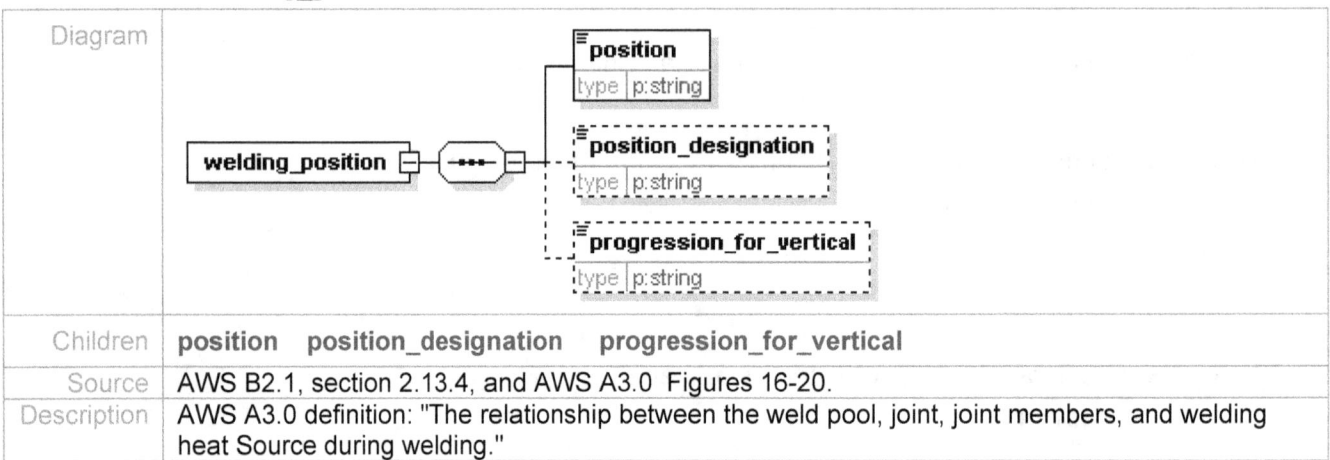
Children	**position position_designation progression_for_vertical**
Source	AWS B2.1, section 2.13.4, and AWS A3.0 Figures 16-20.
Description	AWS A3.0 definition: "The relationship between the weld pool, joint, joint members, and welding heat Source during welding."

*2.19.1 element welding_position/**position***

Type	enumeration of p:string
Values	enumeration flat enumeration horizontal enumeration vertical enumeration overhead enumeration all_positions(for PQRs)
Description	Name of the position.

*2.19.2 element welding_position/**position_designation***

Type	enumeration of p:string
Values	enumeration 1F enumeration 1G enumeration 2F enumeration 2G enumeration 4F enumeration 4G enumeration 5F enumeration 5G
Source	AWS A3.0, Figures 16-20.
Description	The two-character abbreviation for the position.

2.19.3 element welding_position/*progression_for_vertical*

Type	enumeration of p:string
Values	enumeration uphill enumeration downhill
Source	AWS B2.1, section 2.13.4 (2).
Description	Direction of torch travel on a weld that is inclined away from the horizontal position.

2.20 element **welding_procedure_specification**

Diagram	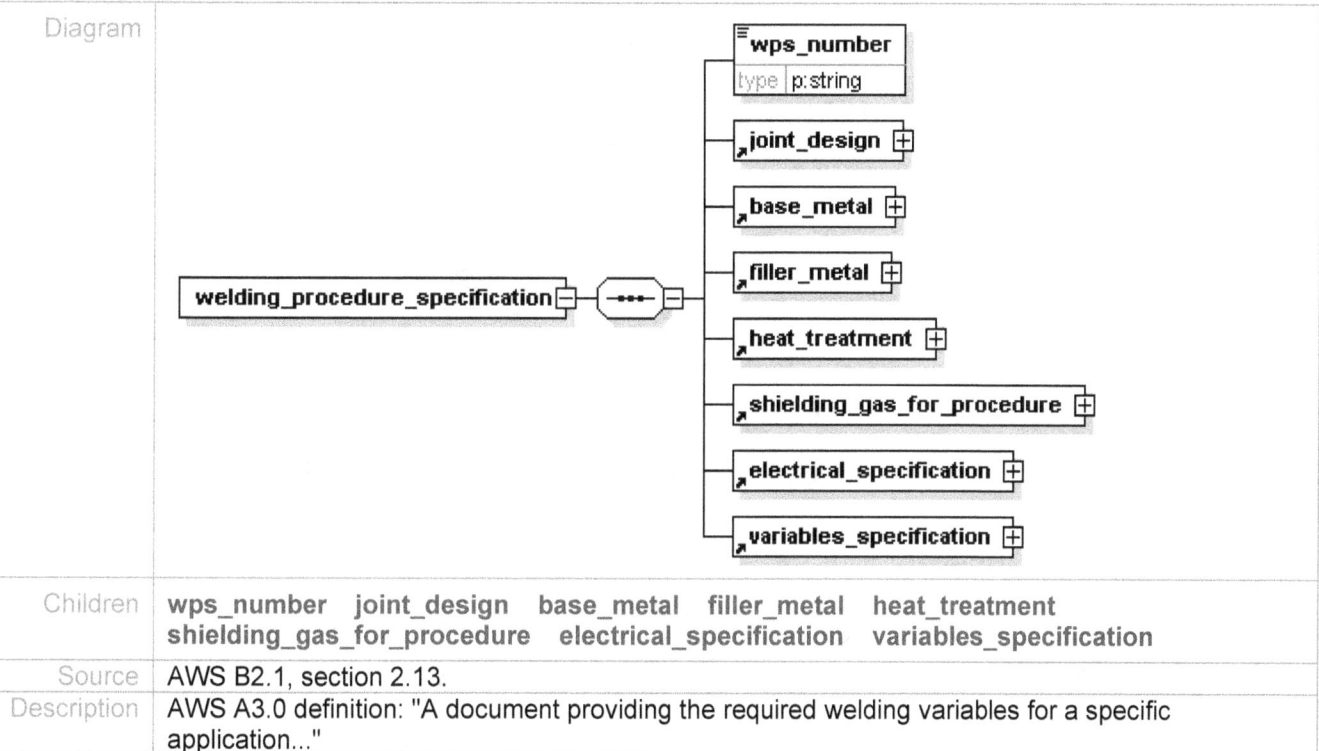
Children	**wps_number joint_design base_metal filler_metal heat_treatment shielding_gas_for_procedure electrical_specification variables_specification**
Source	AWS B2.1, section 2.13.
Description	AWS A3.0 definition: "A document providing the required welding variables for a specific application..."

2.20.1 element welding_procedure_specification/*wps_number*

Type	p:string
Source	Use of WPS number appears in many suggested report forms, including AASHTO/AWS D1.5M:D1.5/2002, Form III-2, and AWS B4.0, Figure E14.
Description	A unique alphanumeric identifier for a specific WPS.

2.21 type welding_process_type

Diagram	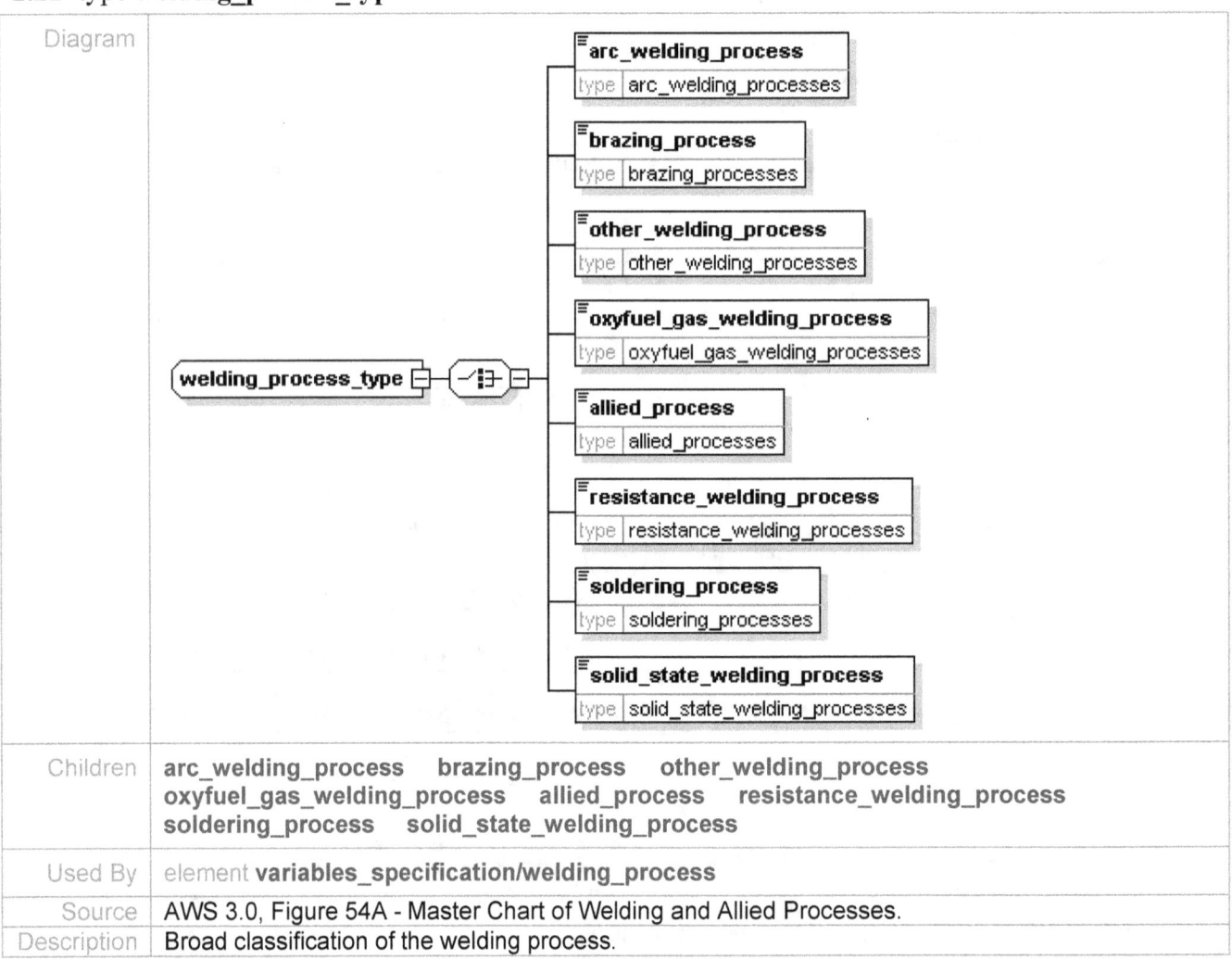
Children	arc_welding_process brazing_process other_welding_process oxyfuel_gas_welding_process allied_process resistance_welding_process soldering_process solid_state_welding_process
Used By	element **variables_specification/welding_process**
Source	AWS 3.0, Figure 54A - Master Chart of Welding and Allied Processes.
Description	Broad classification of the welding process.

2.21.1 element welding_process_type/*arc_welding_process*

Type	arc_welding_processes	
Values	enumeration	atomicHydrogenWelding(AHW)
	enumeration	bareMetalArcWelding(BMAW)
	enumeration	carbonArcWelding(CAW)
	enumeration	carbonArcWeldingGas(CAW-G)
	enumeration	carbonArcWeldingShielded(CAW-S)
	enumeration	electrogasWelding(EGW)
	enumeration	electroSlagWelding(ESW)
	enumeration	gasMetalArcWelding(GMAW)
	enumeration	gasTungstenArcWelding(GTAW)
	enumeration	plasmaArcWelding(PAW)
	enumeration	shieldedMetalArcWelding(SMAW)
	enumeration	studArcWelding(SW)
	enumeration	submergedArcWelding(SAW)
	enumeration	submergedArcWeldingSeries(SAW-S)

Description	See AWS 3.0, Figure 54A - Master Chart of Welding and Allied Processes

2.21.2 element welding_process_type/*brazing_process*

Type	brazing_processes	
Values	enumeration	block_brazing(BB)
	enumeration	diffusion_brazing(CAB)
	enumeration	dip_brazing(DB)
	enumeration	exothermic_brazing(EXB)
	enumeration	flow_brazing(FLOW)
	enumeration	furnace_brazing(FB)
	enumeration	induction_brazing(IB)
	enumeration	infrared_brazing(IRB)
	enumeration	resistance_brazing(RB)
	enumeration	torch_brazing(TB)
	enumeration	twin_carbon_arc_brazing(TCAB)
Description	See AWS 3.0-94 p. 100, 101	

2.21.3 element welding_process_type/*other_welding_process*

Type	other_welding_processes	
Values	enumeration	electron_beam_welding(EBW)
	enumeration	electroslag_welding(ESW)
	enumeration	flow_welding(FLOW)
	enumeration	induction_welding(IW)
	enumeration	laser_beam_welding(LBW)
	enumeration	percussion_welding(PEW)
	enumeration	thermite_welding(TW)
Description	See AWS 3.0, Figure 54A - Master Chart of Welding and Allied Processes	

2.21.4 element welding_process_type/*oxyfuel_gas_welding_process*

Type	oxyfuel_gas_welding_processes	
Values	enumeration	air_acetylene_welding(AAW)
	enumeration	oxyacetylene_welding(OAW)
	enumeration	oxyhydrogen_welding(OHW)
	enumeration	pressure_gas_welding(PGW)
Description	See AWS 3.0, Figure 54A – Master Chart of Welding and Allied Processes	

2.21.5 element welding_process_type/*allied_process*

Type	allied_processes	
Values	enumeration	oxygen_cutting(OC)
	enumeration	arc_cutting(AC)
	enumeration	other_cutting
Description	See AWS 3.0, Figure 54A - Master Chart of Welding and Allied Processes	

*2.21.6 element welding_process_type/**resistance_welding_process***

Type	resistance_welding_processes	
Values	enumeration	flash_welding(FW)
	enumeration	projection_welding(PW)
	enumeration	resistance_spot_welding(RSW)
	enumeration	resistance_weam_welding(RSEW)
	enumeration	upset_welding(UW)
Description	See AWS 3.0, Figure 54A - Master Chart of Welding and Allied Processes	

*2.21.7 element welding_process_type/**soldering_process***

Type	soldering_processes
Description	See AWS 3.0, Figure 54A - Master Chart of Welding and Allied Processes

*2.21.8 element welding_process_type/**solid_state_welding_process***

Type	solid_state_welding_processes	
Values	enumeration	coextrusion_welding(CEW)
	enumeration	cold_welding(CW)
	enumeration	diffusion_welding(DFW)
	enumeration	explosion_welding(EXW)
	enumeration	forge_welding(FOW)
	enumeration	friction_welding(FRW)
	enumeration	hot_pressure_welding(HPW)
	enumeration	ultrasonic_welding(USW)
Description	See AWS 3.0, Figure 54A - Master Chart of Welding and Allied Processes	

3. Testing and Inspection Data

Terms typically used in inspection applications that already appear in the previous sections will not be duplicated in this section. The scope is data that is essential to reporting results of tests. In addition, some fields were added that are found on test results forms that are essential for record keeping and for tracing the source of results. Example data are testing machine type, model and serial number, technician running the test, and date of test.

The scope of this data dictionary addresses mechanical testing, but not non-destructive (e.g., Welding Handbook, Vol. 1, p. 213), metallurgical or compositional testing. These areas need to be documented for complete coverage of weld testing procedures and their results.

3.1 element **bend_test_bend_radius**

Diagram	
	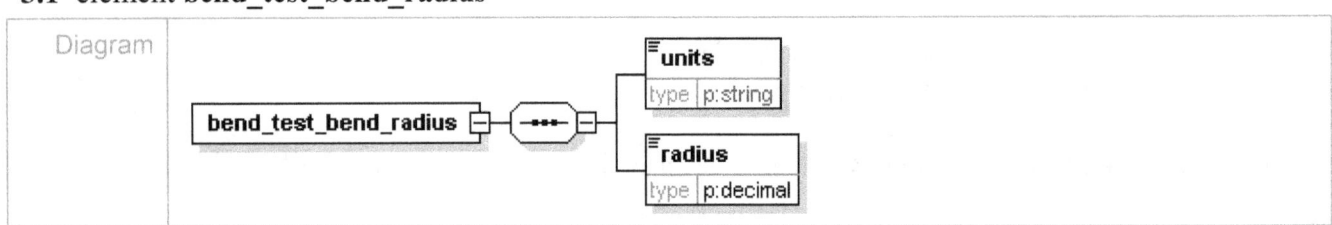

Children	**units** radius
Source	AWS B4.0, section A1-9(6)
Description	Radius of the plunger or mandrel used in a bend test.

3.1.1 element bend_test_bend_radius/*units*

Type	enumeration of p:string
Values	enumeration millimeters enumeration inches
Description	Choice of U.S. Customary or SI units.

3.1.2 element bend_test_bend_radius/*radius*

Type	linear_dimension_type_mm
Description	The numerical value.

3.2 type **bend_test_types**

Type	enumeration of p:string
Values	enumeration transverse_side_bend enumeration transverse_face_bend enumeration transverse_root_bend enumeration longitudinal_face_bend enumeration longitudinal_root_bend enumeration fillet_weld_root_bend enumeration surfacing_weld_face_bend enumeration surfacing_weld_side_bend
Source	AWS B4.0, section A1-7
Description	Classification of bend test specified.

3.3 element **fracture_crack_plane_orientation**

Type	p:string
Source	AWS B4.0, section A3-9
Description	Location of the fracture in relation to the weld.

3.4 element **fracture_energy_absorbed**

Diagram	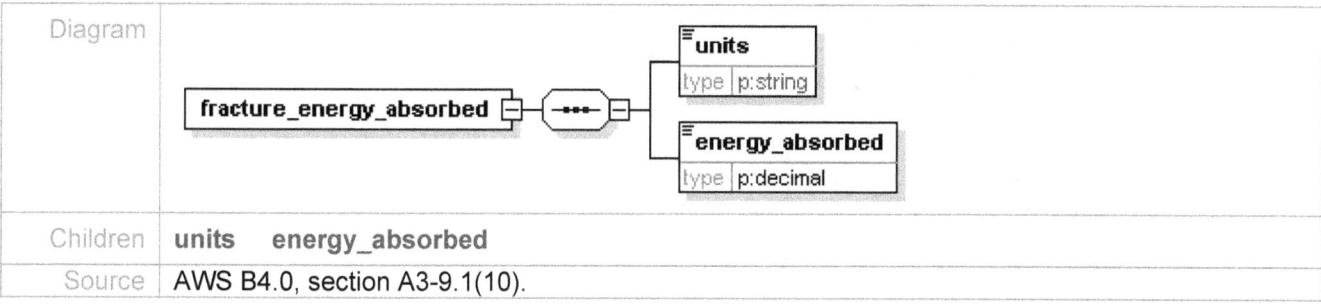
Children	**units** energy_absorbed
Source	AWS B4.0, section A3-9.1(10).

| Description | In a fracture toughness test, the energy consumed in breaking the specimen. |

3.4.1 element fracture_energy_absorbed/*units*

Diagram	
Type	enumeration of p:string
Values	enumeration joules enumeration foot_pounds
Description	Choice of U.S. Customary or SI units.

3.4.2 element fracture_energy_absorbed/*energy_absorbed*

Type	p:decimal
Description	The value of the test result.

3.5 element **fracture_machine_notch_position**

Type	p:string
Source	AWS B4.0, section A3-9
Description	Location of the machined notch.

3.6 element **fracture_specimen_location**

Type	p:string
Source	AWS B4.0, section A3-9
Description	Area of the weldment the specimen was cut from.

3.7 type **fracture_toughness_test_methods**

Type	enumeration of p:string
Values	enumeration charpy_v_notch_E_23 enumeration dynamic_tear_E_604 enumeration plane_strain_E_399 enumeration drop_weight_nil_ductility_temp_E_208
Source	AWS B4.0, section A3.1
Description	Name of the specific ASTM test specified.

3.8 type **fracture_type_of_test_equipment**

Type	p:string
Source	AWS B4.0, section A3-9

| Description | Mechanical configuration of test machine used, to conform to a specified test method. |

3.9 element **hardness_indentor**

Type	p:string
Source	AWS B4.0, section C2-9(7)
Description	Specification for the indentor used in the test. This is one data element of a test report.

3.10 element **hardness_load**

Type	p:decimal
Source	AWS B4.0, section C2-9(7)
Description	The force load used in the test, units must be specified. This is one data element of a test report.

3.11 element **hardness_location_of_impressions**

Type	p:string
Source	AWS B4.0, section C2-9(8)
Description	Description of locations of multiple test impressions.

3.12 element **hardness_test_result**

Source	AWS B4.0, section C219.(10).
Description	The numerical hardness value obtained by the test.

3.13 element **hardness_test_types**

Type	enumeration of p:string	
Values	enumeration	brinell_E_10
	enumeration	rockwell_E_18
	enumeration	vickers_E_92
	enumeration	Knoop
	enumeration	vickers_E_384
	enumeration	portable_hardness_E_110
	enumeration	other_rockwell
Source	AWS B4.0, section C2-4	
Description	The name of the test specified or required.	

3.14 element mechanical_test

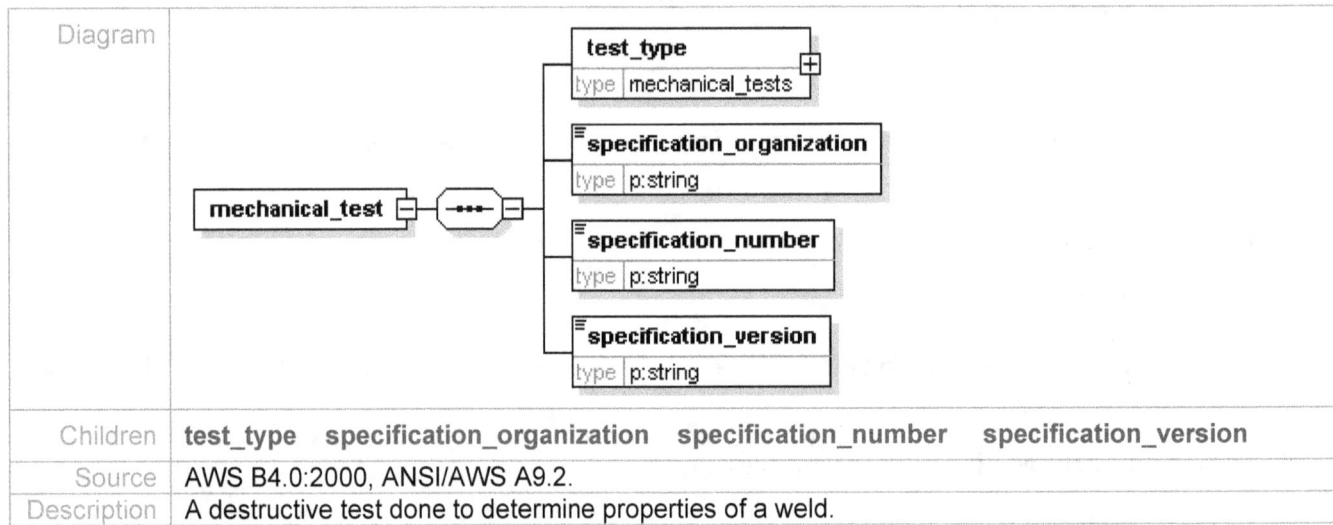

Children	test_type specification_organization specification_number specification_version
Source	AWS B4.0:2000, ANSI/AWS A9.2.
Description	A destructive test done to determine properties of a weld.

3.14.1 element mechanical_test/test_type

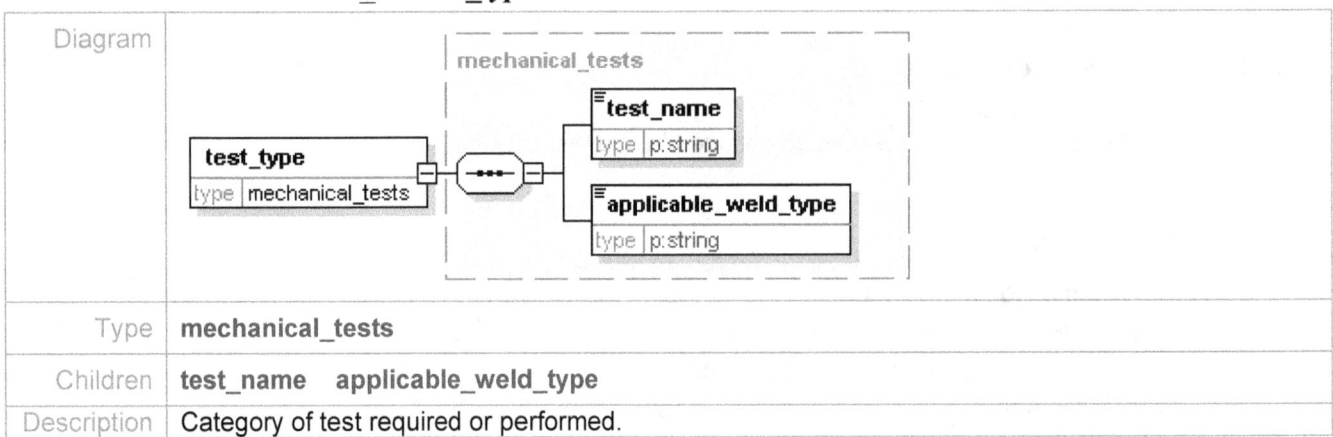

Type	mechanical_tests
Children	test_name applicable_weld_type
Description	Category of test required or performed.

3.14.2 element mechanical_test/specification_organization

Type	p:string
Source	ANSI/AWS A9.2. sections 5.1.1.1, 5.2.1.1, 5.3.1.1

3.14.3 element mechanical_test/specification_number

Type	p:string
Source	ANSI/AWS A9.2. sections 5.1.1.2, 5.2.1.2, 5.3.1.2

3.14.4 element mechanical_test/specification_version

Type	p:string

Source	ANSI/AWS A9.2. sections 5.1.1.3, 5.2.1.3, 5.3.1.3
Description	The version of the test method used.

3.15 type **mechanical_tests**

Diagram	(mechanical_tests diagram showing test_name (type p:string) and applicable_weld_type (type p:string))
Children	**test_name** **applicable_weld_type**
Used By	element **mechanical_test/test_type**
Source	AWS B4.0:2000
Description	Name of test type and the weld type being evaluated.

3.15.1 element mechanical_tests/*test_name*

Type	enumeration of p:string	
Values	enumeration	bend
	enumeration	face_bend
	enumeration	root_bend
	enumeration	side_bend
	enumeration	tension
	enumeration	fracture_toughness
	enumeration	longitudinal_guided_bend
	enumeration	soundness
	enumeration	shear
	enumeration	nick_break
	enumeration	hardness
	enumeration	stud_weld
	enumeration	controlled thermal_severity
	enumeration	cruciform
	enumeration	implant
	enumeration	lehigh_restraint
	enumeration	varestraint
	enumeration	oblique_y_groove
Source	AWS B4.0:2000	

3.15.2 element mechanical_tests/*applicable_weld_type*

Type	enumeration of p:string	
Values	enumeration	groove_weld
	enumeration	fillet_weld
	enumeration	groove_and_fillet_welds

		enumeration	stud_weld
		enumeration	weldability_test
Description	AWS B4.0 classifies tests by weld Type, ie groove, fillet, stud, etc.		

3.16 element **nick_break_apparatus**

Type	enumeration of p:string	
Values	enumeration	fracture_by_hammer
	enumeration	loading_in_tension
	enumeration	three_point_bending
Source	AWS B4.0, section C1-6	
Description	Type of test performed	

3.17 type **shear_specimen_types**

Type	enumeration of p:string	
Values	enumeration	longitudinal
	enumeration	transverse
Source	AWS B4.0, section B3-7	
Description	Orientation of the specimen to that of the sampled weld.	

3.18 element **shear_test_shear_strength**

Diagram	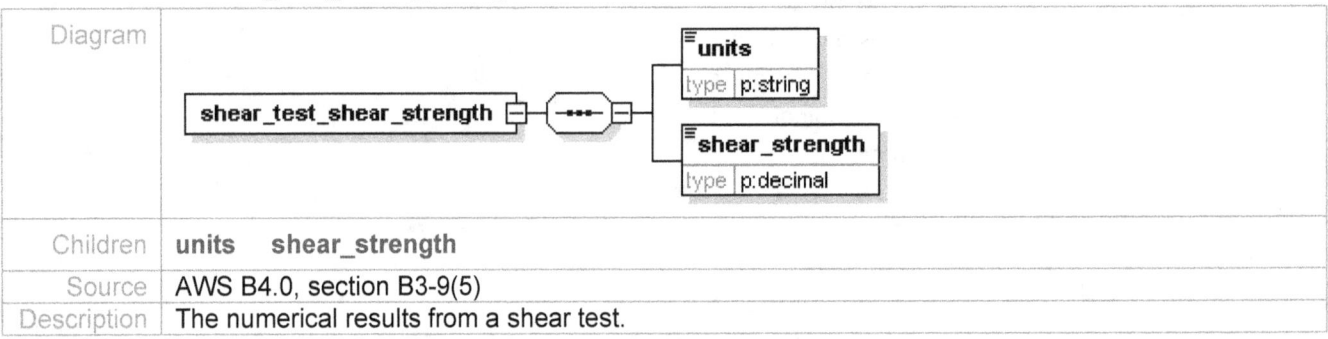
Children	**units shear_strength**
Source	AWS B4.0, section B3-9(5)
Description	The numerical results from a shear test.

3.18.1 element shear_test_shear_strength/**units**

Type	p:string
Description	SI or U.S. customary units.

3.18.2 element shear_test_shear_strength/**shear_strength**

Type	p:decimal
Description	Numerical results of the test.

3.19 element **shear_test_unit_shear_load**

Diagram	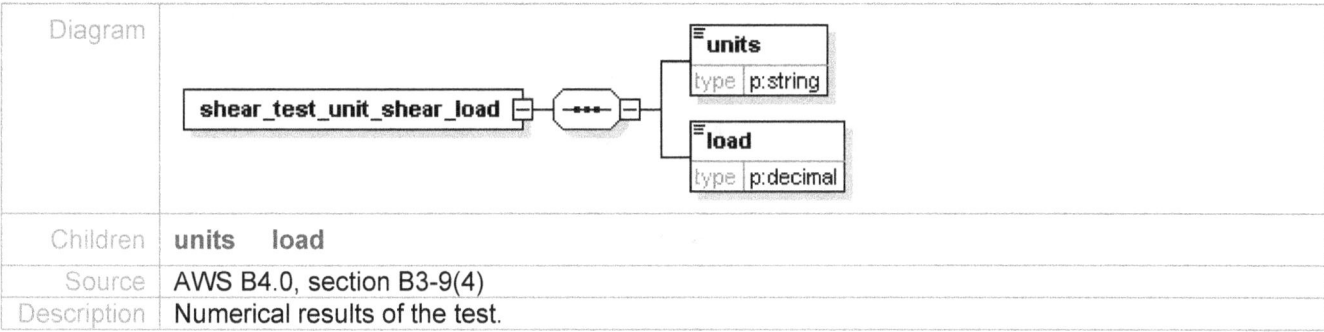
Children	**units** **load**
Source	AWS B4.0, section B3-9(4)
Description	Numerical results of the test.

3.19.1 element *shear_test_unit_shear_load/**units***

Type	p:string
Description	Choice of U.S. Customary or SI system.

3.19.2 element *shear_test_unit_shear_load/**load***

Type	p:decimal
Description	Numerical value of the test result.

3.20 type **soundness_test_types**

Type	enumeration of p:string	
Values	enumeration	fillet_weld_break_proc_qual
	enumeration	fillet_weld_break_primer_coated_p_q
	enumeration	fillet_weld_break_galv_proc_qual
	enumeration	fillet_weld_break_welder_qual
	enumeration	fillet_weld_break_tack_welder_qual
Source	AWS B4.0, section B2-7	
Description	Criterion for using the results of the test.	

3.21 type **tension_specimen_types**

Type	enumeration of p:string	
Values	enumeration	round_all_weld_metal_type
	enumeration	round_transverse_type
	enumeration	rectangular_transverse_type
	enumeration	rectangular_longitudinal_type
	enumeration	tubular_type
Source	AWS B4.0, section A2-7.	
Description	Description of shape of the specimen and where in the weld it was taken.	

3.22 element **tension_test_data**

Diagram	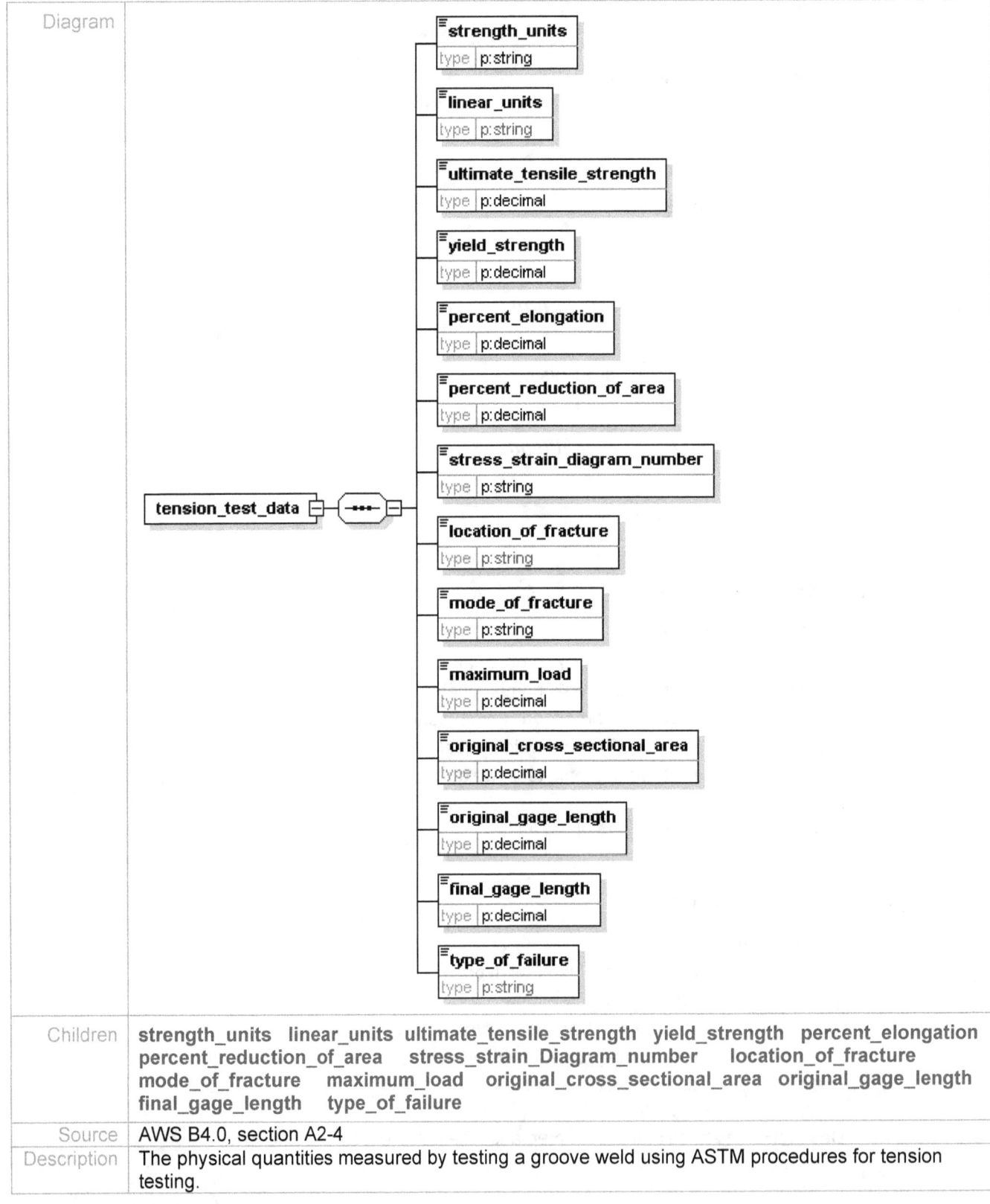
Children	strength_units linear_units ultimate_tensile_strength yield_strength percent_elongation percent_reduction_of_area stress_strain_Diagram_number location_of_fracture mode_of_fracture maximum_load original_cross_sectional_area original_gage_length final_gage_length type_of_failure
Source	AWS B4.0, section A2-4
Description	The physical quantities measured by testing a groove weld using ASTM procedures for tension testing.

3.22.1 element tension_test_data/*strength_units*

Type	Enumeration of p:string
Description	Choice of SI or U.S. Customary units.

3.22.2 element tension_test_data/*linear_units*

Type	enumeration of p:string	
Values	enumeration	millimeters
	enumeration	inches
Description	Choice of SI or U.S. Customary units.	

3.22.3 element tension_test_data/*ultimate_tensile_strength*

Type	p:decimal
Description	Highest engineering stress in the material before rupture.

3.22.4 element tension_test_data/*yield_strength*

Type	p:decimal
Description	Maximum stress that can be induced before the material goes plastic.

3.22.5 element tension_test_data/*percent_elongation*

Type	p:decimal
Description	A measure of ductility, the change in gage length divided by the original length times 100.

3.22.6 element tension_test_data/*percent_reduction_of_area*

Type	p:decimal
Description	Difference between the original cross sectional area and the smallest specimen cross section after testing.

3.22.7 element tension_test_data/*stress_strain_Diagram_number*

Type	p:string
Description	Drawing number of the plot of stress versus strain test results.

3.22.8 element tension_test_data/*location_of_fracture*

Type	enumeration of p:string	
Values	enumeration	unaffected base metal
	enumeration	weld metal
	enumeration	heat affected zone

Description	Type of material in the specimen where the fracture occurred.

3.22.9 element tension_test_data/*mode_of_fracture*

Type	p:string
Description	Mechanics of material deformation that caused the fracture.

3.22.10 element tension_test_data/*maximum_load*

Type	p:decimal
Description	Maximum force exerted on the specimen during the test.

3.22.11 element tension_test_data/*original_cross_sectional_area*

Type	p:decimal
Description	Area of the specimen before testing.

3.22.12 element tension_test_data/*original_gage_length*

Type	p:decimal
Description	Length of the specimen before testing.

3.22.13 element tension_test_data/*final_gage_length*

Type	p:decimal
Description	Specimen length after testing, measured to determine elongation.

3.22.14 element tension_test_data/*type_of_failure*

Type	enumeration of p:string
Values	enumeration ductile enumeration brittle
Source	AWS B4.0, section A2-9
Description	Broad category of failure.

3.23 element **test_fracture_appearance**

Type	p:string
Source	AWS B4.0, section A3-9
Description	Visible evidence of the fracture.

3.24 element **test_fracture_location**

Type	p:string

Source	AWS B4.0, section A3-9.
Description	Area of failure relative to the weld.

3.25 element **test_LGBTFW_angle_of_fracture**

Type	p:string
Source	AWS B4.0, section B1-9(6)
Description	Longtitudinal guided-bend test for fillet welds, angle of fracture.

3.26 element **test_LGBTFW_discontinuity**

Diagram	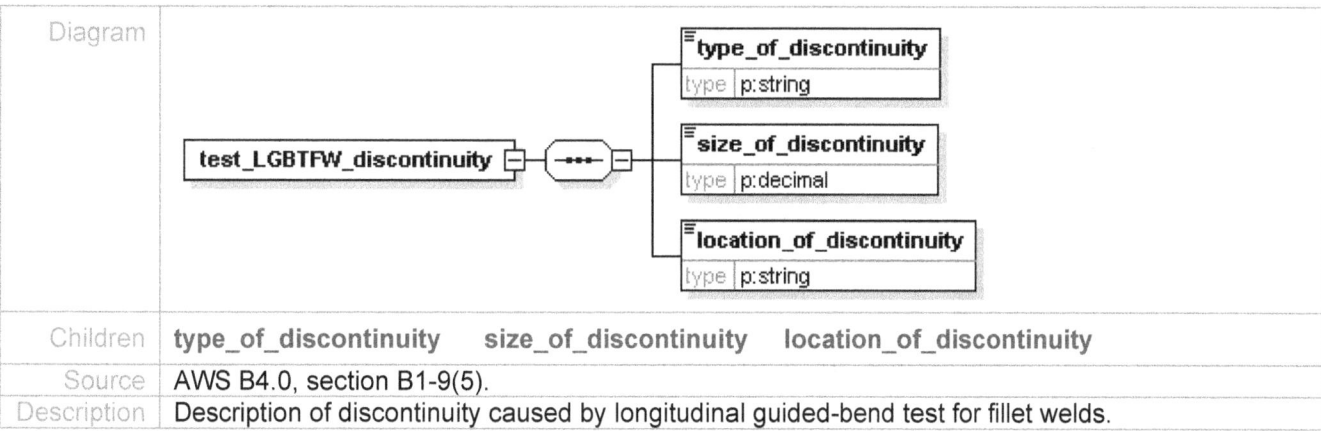
Children	**type_of_discontinuity size_of_discontinuity location_of_discontinuity**
Source	AWS B4.0, section B1-9(5).
Description	Description of discontinuity caused by longitudinal guided-bend test for fillet welds.

3.26.1 element test_LGBTFW_discontinuity/**type_of_discontinuity**

Diagram	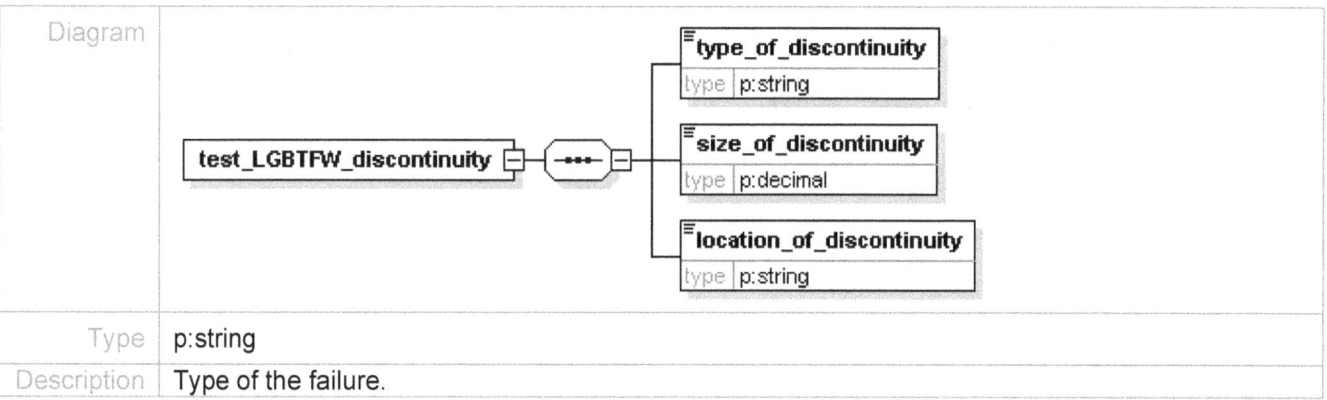
Type	p:string
Description	Type of the failure.

3.26.2 element test_LGBTFW_discontinuity/**size_of_discontinuity**

Type	p:decimal
Description	Editor's note: Is this a quantitative measure or observation of relative size? Need units specified if it is a measurement.

3.26.3 element test_LGBTFW_discontinuity/**location_of_discontinuity**

Type	p:string

| Description | Location in relation to the weld. |

3.27 element test_machine_serial_number

Type	p:string
Description	Serial number to identify the machine that produced the test results.

3.28 element test_number_of_specimens

Type	p:integer
Source	AWS B4.0
Description	The number of specimens of a single weld required for a test.

3.29 element test_percent_elongation

Type	p:decimal
Source	AWS B4.0, section A1, 1.4
Description	In a tension test, the ratio of the change in length to the initial length.

3.30 element test_postweld_mechanical_treatment

Type	p:string
Description	Any physical working of the weld specimen before testing.

3.31 element test_postweld_thermal_treatment

Type	p:string
Source	AWS B4.0, section A1-1.3
Description	Times and temperatures of heat treatment after welding. It is possible this element should be further decomposed if it's information can be structured.

3.32 element test_specimen_dimensions

Diagram	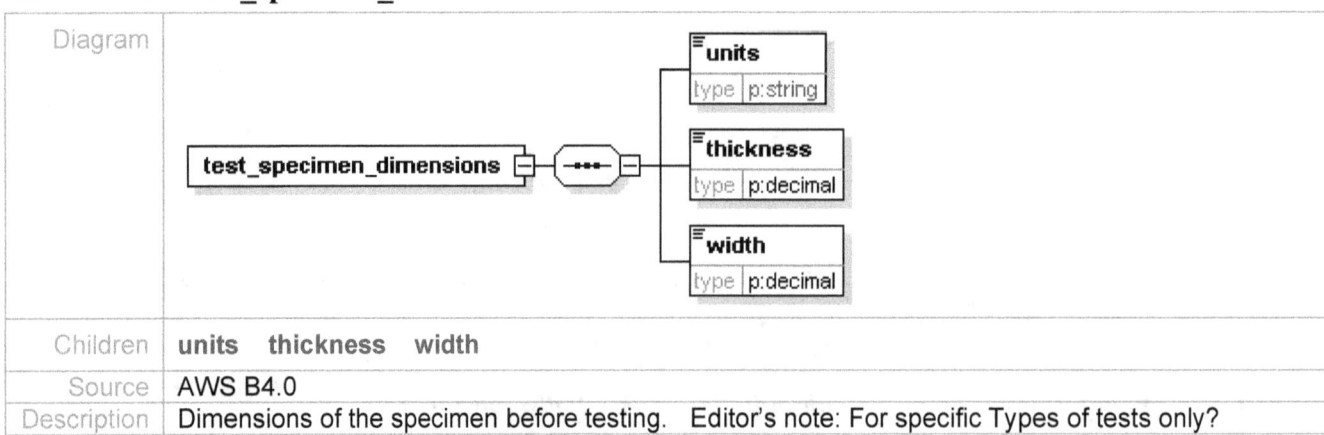
Children	**units** **thickness** **width**
Source	AWS B4.0
Description	Dimensions of the specimen before testing. Editor's note: For specific Types of tests only?

3.32.1 element test_specimen_dimensions/*units*

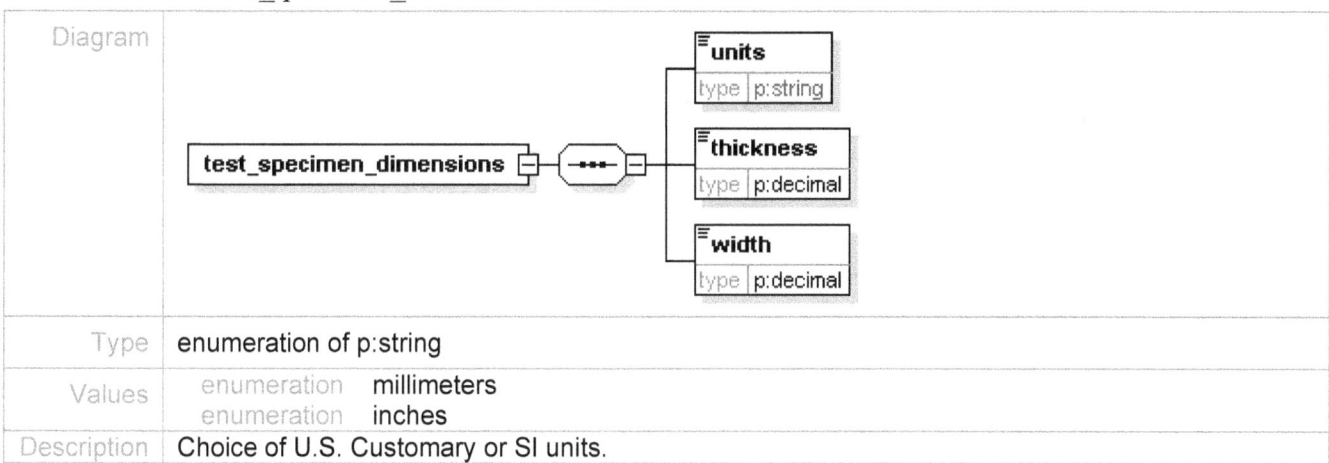

Diagram	
Type	enumeration of p:string
Values	enumeration millimeters enumeration inches
Description	Choice of U.S. Customary or SI units.

3.32.2 element test_specimen_dimensions/*thickness*

Type	p:decimal
Description	Thickness of the specimen before testing.

3.32.3 element test_specimen_dimensions/*width*

Type	p:decimal
Description	Width of the specimen before testing.

3.33 element **test_specimen_location**

Type	p:string
Source	AWS B4.0, section A1, 1.3, 1.4
Description	Where in the weldment/joint/weld the specimen was cut.

3.34 element **test_specimen_orientation**

Type	enumeration of p:string
Values	enumeration longitudinal enumeration transverse
Source	AWS B4.0M, p. 1
Description	The relationship of the axis of the test specimen to the axis of the weld.

3.35 element **test_technician_name**

Type	p:string
Source	Taken from an example of an industry test report.
Description	Name of the person who conducted a test.

3.36 type **test_temperature**

Diagram	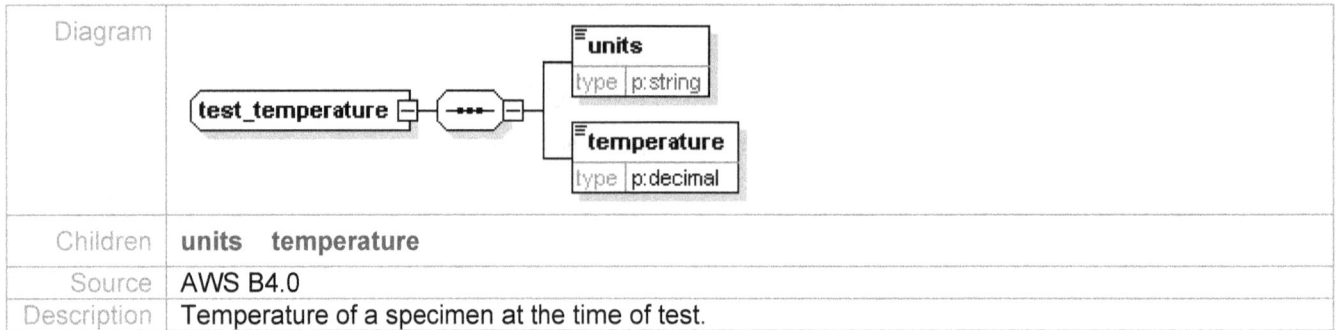
Children	units temperature
Source	AWS B4.0
Description	Temperature of a specimen at the time of test.

3.36.1 element test_temperature/*units*

Type	enumeration of p:string
Values	enumeration degrees_celsius enumeration degrees_fahrenheit
Description	Choice of SI or U.S. Customary units celsius or fahrenheit.

3.36.2 element test_temperature/*temperature*

Type	p:decimal

Acknowledgments

The technical information came primarily from American Welding Society documents that were generated by hundreds of experts in diverse fields. This document is an extension of two 1992 AWS documents, "A9.1" and "A9.2" that were written by the AWS A9 committee then chaired by Jerry E. Jones. Tom Siewert has for some time encouraged the pursuit of a standard data specification for welding data. Tim Quinn had extensive technical input to this document. Christopher Pepper helped edit the document.

Appendices

Appendix 1 – Items to Add in Next Edition of Data Dictionary

Design

- three dimensional geometric location of weld (including weld begin, end points, contour), weld length.
- information to describe the required accuracy of cutting of base materials (e.g., cutting, flame cutting, laser cutting), and accuracy of edge preparation, e.g., whether machining is required, or if flame cutting is adequate.
- tubular joints, e.g., Welding Handbook, Vol. 1, p. 174.

Process
- items needed to describe PQR or WPS ranges, minimum and maximum values, required processes (e.g., "preheat required"), prohibited status (e.g., "consumable inserts are not permitted", or optional conditions.
- non-arc welding process details.

Inspection
- crack types from AWS 3.0 Figure 33, p. 83, and weld discontinuities - Figure 32. These would be used in a visual inspection report.
- welding test positions from AWS 3.0 – Figures 18 and 19.
- non-destructive examination (NDE) parameters and results.
- AWS D14.6, *Specification for Welding of Rotating Elements of Equipment,* is a rich source of welding terms and variables that should eventually be covered.

Note – AWS D14.6

Appendix 2 - Tasks for data modelers

This data dictionary is not sufficiently rich to specify a data base design or a data format. Some of the tasks needed for detailed design are listed below.
- Strengthen relationships of data elements, i.e. grouping elements into reusable chunks and choosing exact encodings of data types and their values (e.g., values for enumerated data). This will be the job of data modelers. They will test the data model for completeness by using scenarios of data flow from industrial welding facilities.
- Harmonize the AWS data model with any that may be developed in specific industries. The "CIMSteel Integration Standard", described at http://www.cis2.org, and http://www.coa.gatech.edu/~aisc/ is an example.
- Specify optional and required data that is dependent on other data – e.g., if base metal is a tube, then "diameter" is required information. For example, if the process is automatic GMAW, gas composition is required data and an electrode size is incorrect data. For example, if "backing_used" is true, then there must be data in "backing_type" – if "backing_used" is false, there cannot be data in the "backing_type" field.
- Allow users to deal easily and unambiguously with U.S. Customary and/or SI units, and also within each of these, a choice of multiple unit scales, e.g., for linear speed, inches/second or feet/minute.
- Decide whether to specify enumerations or to allow free use of text fields. E.g., joint types could be a string field, or an single choice among enumerated values in the specification.
- Specify data types to handle e.g., min/max values of variables in SWPSs and PQRs. Also specify actual ranges that are legal or allowed if appropriate.
- Modularize definitions for maximum reuse in different report forms, and also to reduce the number of elements to aid understandability. For example, many fields from WPSs are used in mechanical test results forms in reporting the processes and weld design that produced the weld.
- Pick encoding options to allow extensibility of data, to allow for conditions not anticipated by the modelers, or to use data unique to a company or a proprietary process. However users are

encouraged to fit their applications within the guidelines of the data models as much as possible, minimizing their use of "other" data types.
- Fine tune definitions and rules for specific standard reports, e.g., mechanical testing results.
- Eliminate duplicate items, in the interest of computational efficiency and understandability.

If an effort goes forward to develop formal data schema, the modelers will cooperate with members of the various AWS committees that authored standards documents to ensure correct interpretation of the data, use of familiar terms, and design of comprehensive and unambiguous enumerations where appropriate.
- Add data elements that welding experts deem required, that are not specified in the AWS documents. E.g., some companies record the serial number of the testing machine used, and the name of the testing technician.
- Pick default values from the enumerated choices (a very detailed decision).

Appendix 3 – The BIG PICTURE of welding data exchange

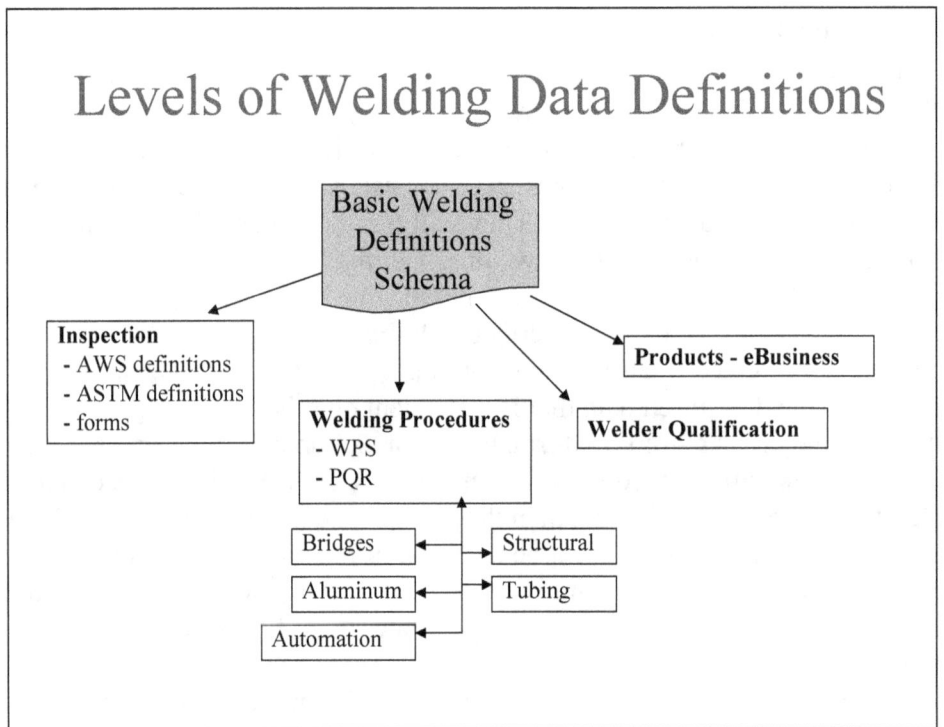

Modularization of welding specifications (instead of pursuing all these areas in a single effort) will speed progress in each smaller area, and allow experts to contribute their knowledge in narrower areas more efficiently.

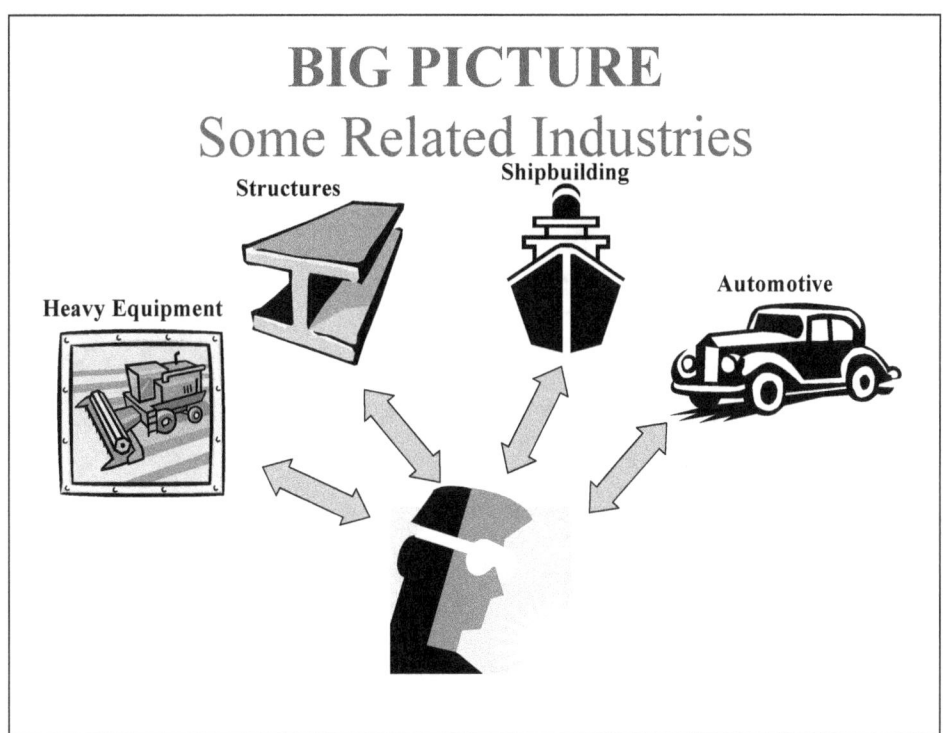

Many manufacturing industries share welding as a fabrication process. The data dictionary helps define welding variables centrally so the industries can "borrow" them, instead of each industry developing their own definitions. The goal is to avoid development of different "dialects" of data formats that the welding industry would have to accommodate.

Appendix 4 – Terminology differences between ISO and AWS
1. ISO does not use the word "flange" in describing joints. It uses "edges turned up".
2. ISO does not describe "groove welds". It uses "butt weld" hyphenated with the name of the groove type.